BIOSPHERE
Forms and Functions

Brief Resume of Dr. B.N. Verma

Dr Bhola Nath Verma was borne on 8th June 1943 and was superannuated from the active service as University Professor, University Department of Botany, B. R. A. Bihar University, Muzaffarpur, Bihar (India). He is the Pioneer Phycologist who initiated and propagated Phycological Research in present Bihar. During thirty-three years of active service, he successfully conducted a number of major research projects sponsored by U.G.C., C.S.I.R., D.S.T., Ministry of Environment and Forests under various ecological as well as Man and Biosphere Programme and Ministry of Non-Conventional Source of Energy, Govt. of India, New Delhi. At least twenty research students were awarded on various aspects of basic as well as applied Phycology and limnology. He supervised C.S.I.R. sponsored post-doctoral research and made a basic contribution on the phenolics of fresh water algae in collaboration with Indian Toxicological Research Institute, Lucknow under the C.S.I.R. scheme for developing Countries granted to one of his research student. He conducted a maiden exploration of edible mushrooms in Arunachal Pradesh. He has published more than hundred research papers including establishment of many new taxa of Green Algae in scholarly journals of National and International repute and authored three reference books. In addition to teaching, research and writings, Prof. Verma actively participated in academic deliberations and chaired a number of technical sessions of National as well as International levels Symposia/ Conferences besides delivering invited lectures. He has been associated as Life Member of seven learned Societies, and as executive of several of them. He is fellow of four Learned Societies. He was conferred Scientist of the year 2001 award by National Environmental Science Academy of India and was felicited by Tribhuvan University, Kathmandu, Nepal in an International Seminar held at its Birganj Campus in 2002 on the recognition of his outstanding contributions in the field of Plant Science. He was elected member of Indo-Nepal Plant Science Forum under the banner of Tribhuvan University. He was honored by Indian Botanical Society in its 26th Annual Conference held in the year 2003 at Zamia Hamdard, New Delhi and by Phycological Society of India in a National Symposium on Biology and Biodiversity of fresh water algae held in the year 2004 at University of Madras, Chennai.

Publications

1. Verma, B.N., 1966. Cytological studies on four species of *Rhizoclonium* Kuetz. *J.R.U.*, 3: 57–63.

2. Verma, B.N., 1967–68. Chromosome number in four Indian species of *Cladophora* Kuetz. *J.R.U.*, 4–5: 61–64.

3. Verma, B.N., 1968. Artificial induction of polyploidy in *Rhizoclonium hieroglyphicum* (Ag) Kuetz. *var. macromeres. Indian Jour. Sc., India*, 2(2): 99–102.

4. Sinha, J.P. and Verma, B.N., 1968. On the occurrence of amitotic nuclear division in *Pithophora cleveana* Wittr. *Phycos*, 7(1 and 2): 139–143.

5. Sinha, J.P. and Verma, B.N., 1969–70. Effect of growth promoting substances on the rate of nuclear division in *Rhizoclonium hieroglyphicum* (Ag) Kuetz. *J.R.U.*, 6–7.

6. Sinha, J.P. and Verma, B.N., 1970. Cytological analysis of the Charophytes of Bihar. *Phycos*, 9(2): 92–99.

7. Verma B.N. and Verma, S.K., 1976. A note on the cytology and life history of *Chaetomorpha gracilis* Kuetz. *Phycos*, 15(1 and 2): 89–94.

8. Verma, B.N. and Verma, M.P., 1977. A new record of *Oedocladium prescottii* Islam from Bihar. *Phycos*, 16(1 and 2): 51–53.

9. Verma, B.N., 1978. New record of *Dichotomosiphon tuberosus* (A. Br.) Ernst *var. indicum var. nov* from Bihar. *Phycos*, 17(1 and 2): 59–62.

10. Verma, B.N., 1979. Cytological studies in four species of *Pithophora* Wittr. *Cytologia*, 44: 29–38.

11. Verma, B.N., 1980. Karyotype analysis in three species of *Rhizoclonium* Kuetz. *Cytologia*, 45: 433–440.

12. Verma, B.N., 1980. Analysis of induced nuclear division in *Rhizoclonium hieroglyphicum* (Ag) Kuetz. *var. macromeres. Acta Botanica Indica*, 8: 215–218.

13. Verma, B.N., 1981. New record of *Cladophora* Kuetz. From Bihar. *Phycos*, 20(1 and 2): 44–48.

14. Verma, B.N., 1982. Karyotype analysis in three species of *Cladophora* Kuetz. *Cytologia*, 47: 137–145.

15. Dogra, A. and Verma, B.N., 1983. First report of *Gongrosira debaryana* Rabenhorst from India. *Bibliotheca Phycologica, J. Cramer*, 66: 371–372.

16. Verma, B.N. and Kumar, A., 1983. First report of chromosome number in *Vaucheria hamata* Wal. *Phycos*, 22(1 and 2): 1–3.

17. Labh, L. and Verma, B.N., 1983. *T. jwalai*, a new species of *Tolypella* from India. *Bibliotheca Phycologica, J. Cramer*, 66: 379–388.

18. Sinha, B.P. and Verma, B.N., 1983. A fresh water Florideae new to India. *Bibliotheca Phycologica, J. Cramer,* 66: 373–377.

19. Labh, L. and Verma, B.N., 1984. Cytological studies in three species of *Tolypella. Cytologia,* 49: 717–723.

20. Verma, B.N., 1985. Karyomorphological studies on three species of *Cladophora* Kuetz. *Cytologia,* 50(1): 25–30.

21. Verma, B.N., 1985. Cytology of *Cladophora liniformis* Kuetz. and *C. fracta* (Dillw.). *Cytologia,* 50 (1): 49–53.

22. Labh, L. and Verma, B.N., 1985. New record of chromosome numbers for the Charophytes. *Cytologia,* 50(1): 55–58.

23. Prasad, Prem K., Sinha, R.S.P. and Verma B.N., 1985. *Some Rare Green Algae from Nepal.* National Academy of Science Letters, India, 8(2): 33–34.

24. Labh, L. and Verma, B.N., 1985. First report of chromosome number in four taxa belonging to the genus *Chara* Linn. *Cytologia,* 50(1): 21–24.

25. Prasad, Prem K. and Verma, B.N., 1985. Aneuploid count for *Chara setosa* Klein ex Willd. *Cytologia,* 50(2): 241–245.

26. Verma, B.N., 1986. Cytotaxonomical studies in the genus *Rhizoclonium* Kuetz. *Cytologia,* 51: 177–183.

27. Verma, B.N., 1986. New counts for *Rhizoclonium riparium* Complex. *Cytologia,* 51: 507–512.

28. Labh, L. and Verma, B.N., 1986. New counts for *Chara fibrosa* Complex. *Cytologia,* 51: 185–191.

29. Verma, B.N. and Labh, L., 1986. Karyomorphological studies in the genus *Chara* Linn. *Cytologia,* 51: 501–506.

30. Verma, B.N. and Jha, C. N., 1988. Effect of diethyl sulphate on *Rhizoclonium* Kuetz. *Cytologia,* 53: 283–286.

31. Verma, B.N. and Prasad, Prem K., 1989. On the cytology of *Microthamnion strictissimum* Rabenh. from Muzaffarpur, Bihar. *Phycos,* 28(1 and 2): 251–253.

32. Jha, C.N. and Verma, B.N., 1989. Non nitrogen fixing cyanobacteria as biofertilizer. *Advances in Applied Phycology,* 11: 153–157.

33. Shukla, C.P. and Verma B.N., 1989. Estimation of total intracellular phenolics and their leachates in culture filtrate during different growth phases of *Oscillatoria subbrevis. Environment and Ecology,* 7 (3): 730–732.

34. Shukla, C.P. and Verma, B.N., 1989. Estimation of total intracellular phenolics during different growth phases of *Chroococcus minor* (Kuetz.) Nag. *Biojournal,* 1(2): 111–112.

35. Sahu, K.K. and Verma, B.N., 1990. New records of *Pithophora* Wittr. from Bihar. *Environ. and Ecol.,* 8(3).

36. Kumar, A. and Verma B.N., 1990. Cellular and extracellular nitrogen content of two N_2 fixing cyanobacteria. *Proc. Nat. Symp. Cyanobacterial Nitrogen Fixation,* p. 163–167.

37. Shukla, C.P. and Verma, B.N., 1990. Quantitative screening of intracellular phenolics during different growth phases of *Synechocystis pevalekii* Ercegovic. *Environ. and Ecol.*, 8(3).

38. Sahu, K.K. and Verma, B.N., 1990. Comparative studies on the effect of chloramphenicol on a nitrogen fixing and non-nitrogen fixing cyanobacteria. In: *Proc. Nat. Symp. Cyanobacterial Nitrogen Fixation*, p. 175–182.

39. Shukla, C. P. and Verma, B.N., 1990. Cellular N$_2$ content and leaching with reference to some non heterocystous cyanobactria. In: *Proc. Nat. Symp. Cyanobacterial Nitrogen Fixation*, p. 169–173.

40. Verma, B.N., Sinha, R., Prasad, Prem K. and Das, H.N., 1990. Blue green algae from the paddy fields of North Bihar. *Phykos*, 29(1 and 2): 55–56.

41. Das, H.N. and Verma, B.N., 1990. Persistent effects of phosphamidon on survival, growth and heterocyst frequency of *Nostoc linckia*. *Environ. and Ecol.*, 8(4): 1142–1146.

42. Kumar, A. and Verma, B.N., 1990. Cellular and extracellular nitrogen content of two N$_2$ fixing cyanobacteria. *Phykos*, 29(1 and 2): 57–62.

43. Verma, B.N. and Sahay, A. P., 1991. Algal phenolics: A protocol. In: *Microbes and Environment*, (Eds.) A.B. Prasad and R.S. Bilgrami. Narendra Publishing House, Delhi, p. 51–74.

44. Das, H.N. and Verma, B.N., 1992. Effect of phosphamidon on non N$_2$ fixing blue-green algae, *Synechocystis aquatilis* Sauv. *Phykos*, 31(1 and 2): 151–157.

45. Sahay A.P., Das, P.K. and Verma, B.N., 1992. Studies on the algal flora of Nepal–I, Chlorophyceae. *Geophytology*, 20(2): 155–158.

46. Singh, M.P. and Verma, B.N., 1993. Interactions amongst the abiotic factors of a fresh water lake. *IBC*, 10(1).

47. Sahay, A.P., Das, P.K. and Verma, B.N., 1993. Studies on the algal flora of Nepal–II, Cyanophyceae and Euglenophyceae. *Geophytology*, 23(1): 181–183.

48. Das, H.N. and Verma, B.N., 1993. Limnology of a pond and a lake at Rajnagar (Madhubani), Bihar. *Biojournal*, 5(1 and 2): 41–46.

49. Das, H.N., Prasad, Prem K. and Verma, B.N., 1994. Methods for ascertaining relative spectrum frequency and rejuvenation potency of a soil algae. *Indian Bot. Soc.*, 73: 35–39.

50. Kund, K. and Verma, B.N., 1994. Biological assessment of algal bloom with special reference to phenolics. In: *Advances in Plant Sciences Research*, Vol. I, (Ed.) K.C. Sahni. International Book, Dehradun, India, 1: 165–178.

51. Das, H.N. and Verma, B.N., 1994. Post treatment impact of carbofuron on *N. linckia* and *S. aquatilis*. *National Bull. Env. Sci.*, 12: 26–28.

52. Kund, K. and Verma, B.N., 1994. Study on the phytoplankton diversity indices of a fish pond. *National Bull. Env. Sci.*, 12.

53. Sahay, A.P., Das, P.K. and Verma, B.N., 1995. Changes in the content of phenolic substances during growth of a cyanobacterium *Oscillatoria surviceps var. angusta*. *Phykos,* 34(1 and 2): 9–16.

54. Verma, B.N. and Das, S.N., 1997. Algal flora of Chitwan and Nawalparasi district of Nepal. *Phykos,* 35(1 and 2): 119–128.

55. Verma, B.N., 1998. Charophyta: Morphotaxonomy and cytotaxonomy. In: *Advances in Phycology,* (Eds.) B.N. Verma *et al.* APC Publication Pvt. Ltd., New Delhi, pp. 173–176.

56. Verma, B.N. and Vidyavati, 1999. Cladophorales: An overview over morphotaxonomy *vis-à-vis* cytotaxonomy. In: *Recent Trends in Algal Taxonomy,* (Ed.) Vidyavati. APC Publication Pvt. Ltd., New Delhi, pp. 101–126.

57. Verma, B.N. and Sinha, A.K., 2002. Studies on the interaction among soil algae *in vitro.* In: *Proceedings of International Symposium on Microbial Biotech. for Sustainable Development,* Jabalpur. Scientific Publisher, Jodhpur.

58. Das, S.N., Verma, B.N. and Vishwanathan, P.N., 2003. Environmental degradation and algae. In: *Phycology: Nature and Nurture,* (Eds.) B.N. Verma *et al.* Kalyani Publication, Delhi, pp. 287–299.

59. Prasad, Prem K. and Verma, B.N., 2003. Exploration of BGA as a possible mean of biological control against nematode disease incidence. In: *Phycology: Nature and Nurture,* (Eds.) B.N. Verma *et al.* Kalyani Publication, Delhi, pp. 197–206.

60. Verma, B.N., Singh, M.P. and Labh, L., 2003. Post-treatment effects of EMS on *Chara* Linn. *Phycology: Nature and Nurture,* (Eds.) B.N. Verma *et al.* Kalyani Publication, Delhi, pp. 273–285.

BIOSPHERE
Forms and Functions

A Festschrift
to
Professor B.N. Verma

— Editor —
Prem Kumar Prasad
Principal
B.M College
Rahika
Madhubani, Bihar

2011
DAYA PUBLISHING HOUSE
Delhi - 110 002

© 2011 PREM KUMAR PRASAD (b. 1957–   )
ISBN 9788170359777

Published by	:	**Daya Publishing House** **A Division of** **Astral International Pvt. Ltd.** **– ISO 9001:2008 Certified Company –** 4760-61/23, Ansari Road, Darya Ganj, New Delhi - 110 002 Phone: 23245578, 23244987 Fax: (011) 23260116 e-mail : dayabooks@vsnl.com website : www.dayabooks.com
Laser Typesetting	:	**Classic Computer Services** Delhi - 110 035
Printed at	:	**Chawla Offset Printers** Delhi - 110 052

PRINTED IN INDIA

Dr. B.N. Verma

Foreword

This book *"Biosphere: Forms and Functions"* contains selected articles on plant activity under anaerobic conditions, reclamation and management of saline soil and alkali soils, over exploitation of reserved forest, plant diversity, ethnomedicinal important pteridophytes, edible mushrooms, biotechnological renaissance in horticultural crops, tissue culture technology, improvement of protein quality and quantity in soyabean, temperature effects on wheat, chromosome behaviour and structure, seed mycoflora, diseases and biochemical disorders due to fungal contamination and plant antimicrobial potentials. A criticism of Sareen and Wadhwa's (1981) paper entitled "Embryological studies in Papilionaceae: The genus *Alysicarpus* neck"–a critical review is most challenging and bold Endeavour. However, this would have been more appropriate if the reply/comments from the authors were appendexed.

The commemoration volume to honor and recognition to Professor Dr. Bhola Nath Verma is a tribute to an eminent and dedicated scientist. This is the least one can do for the life time achievement of a teacher and scientist. The editor of the volume deserves appreciation for their enthusiasm. I wish a long and healthy life to my esteemed friend and colleague Professor Verma.

I am more than confident the present volume should prove an asset to a large scientific community engaged in teaching and research in institutions and the universities.

Dr. Ajit Verma
Amity University
Uttar Pradesh

Message

Professor M.N. Noor

UGC Visiting Professor, Department of Botany,
Ranchi University, Ranchi – 834 008
Email: m_n_noor@yahoo.com
Tel.: 0651-2491689, 9431597668 (M)

Some Glimpses from the Memory Lane

When I go down to the memory lane, I become nostalgic as certain wavelengths flash out automatically on my mental spectrum pertaining to Professor Bhola Nath Verma with whom I have been emotionally attached for over 4 decades from now. I shall try to catch up just a few flashes of the past golden moments in this brief write-up expecting the readers to kindly bear with me for a little bit historical backdrop of the laboratory where Prof. Verma initiated his scientific career in 1965 under the generous supervision of legendary Phycologist, Prof. Jwala Prasad Sinha.

After completing doctoral work on Cytology of Cladophorales and Oedogoniales in 1958 under the guidance of Professor M.B.E. Godward (London University), Prof. Sinha returned back to Patna University and finally shifted to *Ranchi University* in the year 1961. At that time, research activity in the Department of Botany was a far cry and the scientific infrastructure was virtually dismal. Bestowed with an ardent visionary aptitude, Prof. Sinha established Phycology Laboratory with full facilities in the Department which eventually emerged out as one of the Centres of Phycological Research in the country. In view of his substantial contributions in the field of algal studies., he may rightly be called as *'Father of Algology'* in erstwhile Bihar State of India.

With the facilities available at hand, Prof. Verma joined the laboratory and started working on 'Cytology of Cladophorales'. I have personally seen him virtually playing with chromosomes and I do perfectly recollect his musical vibrations whenever he prepared a well-quashed metaphasic plate in the lab. The moment was really exciting. Basically, he is an Algal Cytologist and his articles, published in reputed Journals, have been well received by the scientific community. I must congratulate him for his work on algal cytology which, my opinion, is quite a difficult proposition.

In the end, I do hope that this Commemoration Volume with around 32 articles published in 2 volumes in honour of Prof. Verma would be interesting and useful to Plant Scientists.

Message

A. Vaishampayan

Professor of Microbial Genetics and Biological Nitrogen Fixation,
B.H.U., Banaras, U.P.
Phone: 0542-6702542/2575358; +91-9415201138;
E-mail: vaishampavan geneticist@yahoo.co.in
prfssr.vaishampavan@gmail.com

Dear Dr. Prem Kumar,

I share my brief views about the academic accomplishments of Prof. B.N. Verma, as follows:

"I am pleased to place on record the glory of my profound academic associations with Professor B.N. Verma from as back as 1969,(when he was an Associate Lecturer at L.S. College Muzaffarpur and I was an I.Sc. student), till date, when he is not in an active service but I have travelled to my present position by virtue of the key of science enlightened in my mind as one of his favourite undergraduate and postgraduate disciples. In between, apart from being one of the most popular and painstaking teacher (Lecturer, and later as a Reader and Professor) of Phycology under the Bihar University services, Professor Verma was ever regarded as an excellent and innovative, researcher on the cytochemical, physico-chemical and biochemical studies of algae with the production of a large number of quality Ph.D. thesis and research publications in referred scientific journals with the receipt of due recognitions far and wide. I was always amazed to find him talking sweet tounged solely in and around the subject of research and nothing else, proving that adding to the knowledge jewel is the greatest earning of life. I am sure this chain of meaningful scientific addition will continue through his brilliant students and students' disciples for time immemorial to keep his name as a noted scholar of Botany, in general, and Algology, in particular. I wish for his long, healthy and active life."

Message

Dr. (Mrs.) Pushpa Srivastava

Emeritus Fellow–UGC,
Algal Biotechnology Laboratory,
Department of Botany,
University of Rajasthan, Jaipur

Prof. B.N. Verma of Department of Botany, University of Muzaffarpur (Bihar) is a close associate of mine. My association dates back to 1999, when he was delivering his invited lecture in the Conference organised in the honour of Prof. G.S. Venkataraman of Indian Agricultural Research Institute (PUSA) to mark his superannuation. Prof. Verma is a doyen of phycology and was engaged actively in his chosen field–The blue green algae, so dear to his heart. In addition to algae he has equal fascination for higher plants. His collection of plants including a wide variety of cacti stand a witness to his overall interest in plants as a whole. His publications have found a place in National and International Journals both. More than his contributions in the field of phycology, he is known for his courteous nature, open heartedness and ever smiling personality. In person, I have enjoyed his brotherly affection, timely advice and encouragement in the moments of pain and pleasure both. To be a scientist is different from being a noble person, Dr. Verma combines both these God given qualities. If it is true that behind every man there is a women, it is very much true for Mrs. Krishna Verma–simple, affectionate caring and sharing, ever smiling and back bone of her husband. I wish a very healthy prosperous and long life to the couple.

Message

Prof. N. Anand, D.Sc.,

Chief Editor, PHYKOS,
Vice Chancellor, Vels University
Pallavaram, Chennai – 600 117

Dear Dr. Prem,

   I am very much delighted to know that a Commemoration Volume is being brought out in honour of my friend and senior Phycologist of our country Prof. B.N. Verma. Prof. Verma was well known to my teacher Prof. T.V. Desikachary, and the beginning of our association dates back to 1970's at the CAS in Botany, University of Madras, Chennai. Prof. Verma's contribution to phycological research especially cytological studies on algae are significant. His keen interest in the Biodiversity of green algae has revealed the presence of a number of new green algal taxa. He has also been one of the leading Limnologists of the country. I have known Prof. Verma earlier through his publications and later during phycological meetings. He has acquired the distinction of being a great teacher and a serious researcher. I congratulate the authors of the volume and Dr. Prem Prasad for their efforts in compiling the Commemorative publications.

Campus: Velan Nagar, P. V. Vaithiyalingam Road, Pallavaram, Chennai - 600 117, India
Phone: (91-44) 2266 2500/01/02/03 Fax: (91-44) 22662513 Admn. Office: 521/2, Anna Salai.
(Opp. G.R. Complex), Nandanam, Chennai - 600 035. Phone: 2431 5541 /42 Website:
www.velsuniv.org E-mail: velscollege@gmail.com.

Preface

While celebrating superannuating of our revered teacher and guide Prof. B. N. Verma, sentiments and emotions broke down. An idea flashed down at the moment to present something special to our guru which could be memorial, a token of respect and befitting to his scientific temperament. Publication of a volume in his honor/ dedicated to him was instantly considered the only way to mark our respect to this mentor. The drive was on to collect contributions from contemporary phycologists with long working experience in the respective domain. It was a difficult time for the editor to select out of the contributions for the comprehensive volume. Participation of some close associates of Prof. Verma, however, could not be ignored. It was this reason that the conservative approach on Algae had to be dropped and a broader perspective was undertaken to accommodate these contributions.

The present compendium contains altogether nineteen research articles of which eleven are devoted to basic and fundamental aspects of Cyanobacteria and Algae. The biodiversity status of Chilka lake, algal flora in polluted environments and the botanical ramble among members of Cyanobacteria in natural and artificial symbioses speak of an in depth observations of expert in the field. The systematic study and phylogenetic treatment of the mixotrophic Euglenineae has also been discussed. Besides these topics of fundamental interest, biotechnological promises, soil reclamation and some experimental works concerning nitrogen metabolism, antibiotic and radiation effect of Cyanobacteria find a place in this volume. Remaining eight articles are concerned with higher plants and heterotrophic fungi. The ethno medicinal importance of some common vascular Cryptogams, the Pteridophytes, in the Aravalli range of Rajasthan have been discussed. Some horticultural crops and their biotechnological renaissance have been looked into. Spices cultivated in India and their known anti-microbial activities have been enumerated. Developing weevil resistance in an economic plant like Sweet Potato through tissue culture technique is an important achievement in biotechnology. This has been discussed at length. The seedling diseases due to storage fungi of the crop seeds, the associated metabolic disorder and the study of mycoflora in plants belonging to Umbelliferae have been well described. The structure and behavior of eucaryotic chromosomes in Radish and over exploitation of Victoria Park in Gujarat causing the degradation of the forest are some of the contributions in the present volume.

The present treatise is thus hetero-organic ranging from procaryotic algae to Vascular Cryptogams to higher plants on one hand and on the other from mixotrophic forms to heterotrophs.

Editor is indeed grateful to authors and experts in the field for generously contributing their papers and thoughtfully extending all out co-operation, which made this effort a success.

Prem Kumar Prasad

Contents

Chapter 1

Plant Activity in Anaerobic Environment Under Water Submergence in Wetlands

Ashwani K. Srivastava

Dean, Faculty of Basic Sciences and Humanities,
Rajendra Agricultural University,
Pusa – 848 125, Samastimpur, Bihar

In India the Green Revolution had tended to concentrate in certain well–endowed pockets, and these have made an enormous contribution to the Nation's food supply. Between 1951 tp 1991 India's population grew by 134 percent, from 361 million to 846 million, but cereal production increased only by 229 percent, from 54 to 174 million metric tons. The per capita production rose from 150 to 206 kilograms per year, an average annual increase of about 11 percent. Most of this production gain came from states in which the Green Revolution had its greatest impact. Now the major challenges of food production are with the states which have yet to use Green Revolution after removing the resource constraints and developing will to eradicate poverty and hunger simultaneously protecting the natural resource base. Looking to the statistics of world population growth, there is an unprecedented increase by 90 million people a year in the next quarter century, has developed a dangerous sense of complacency about the current and future food, agriculture, and environment situation. About 800 million people, 20 percent of the developing world's population were food insecure in 1995, the majority lack economic and physical access to the food required to lead healthy and productive lives. Hence in the next 25 years, the world will be challenged to produce enough food to feed an additional 90 million people each year, as well to meet increasing and changing food needs due to rising incomes and changing life styles. So by 2020 there needs to develop technology for growing crops under fragile environment to exploit to the maximum the land potentiality.

Drought and water submergence are the two unpredictable main environmental constraints which affect crop growth and yield. Water submergence at critical plant growth stage proves a serious threat to agriculture in major part of south Asia. Water logging and submergence occur over vast areas, these include agricultural lands and areas important to the environment such as wetlands and coastal marshes (Kozlowaski, 1984). Submergence occurs in extensive areas called wetlands. There are substantial differences in water logging tolerance between spacies. The general view is that dicotyledons are, usually, more intolerant to water logging than Monocotyledons. The most conspicuous example of a water logging tolerant crop is rice. This species has genotypes which can produce in regions where the crop is prone to submergence. Yet, despite the earlier mentioned high tolerance of rice to water logging, this crop can be adversely affected in survival, growth and grain yield during transient or long term submergence. Large areas of rainfed rice may be adversely affected by sudden flooding, which will result in partial or complete submergence of the shoots.

Water logging is a serious problem in the Indo-Gangetic Basin particularly in the states of Uttar Pradesh, Bihar, West Bengal, Assam, Orissa which causes a great loss in terms of poor crop growth and soil erosion. Water logging in itself induces disorders in plants which became more serious when gets combined with few days of high temperature and light wind velocity, which more often causes temporary wilting in plants (Lie, 1984). Water logging causes a gradual depletion of soil oxygen resulting into toxic elements (Keeley and Franz, 1979). This gradually builds up a condition in plants if remain submerged beyond their tolerance limit, an altered metabolism, inhibiting growth, stomatal closure, reduced photosynthesis and nutrient and water absorption and altered hormone balance (Kozlowoski, 1984).

Depending on types of submergence and genotypes with different tolerance to submergence, the responses can be summarized as follows:

1. Reduction in grain yield due to commitment of carbohydrates to elongation of internodes and leaf sheaths (deep water rices).
2. Cessation of growth in submergent tolerant lowland rice for the period of submergence (transient submergence).
3. Severe damage of foliage and death in submergent intolerant lowland rices.

One of the most important factor which affects plant growth either in waterlogging or submergence is the slow gas diffusion. During both waterlogging and submergence the 10,000 fold slower diffusion of gases in solution than in the gas phase is almost certainly one of the most important factors determining plant response. The slow gas diffusion results in depletion of gases in soil and flood waters which are absorbed by the plants and micro-organisms and accumulation of gases which are produced by the plants or micro-organisms. For flood waters with submerged foliage this general rule gives rise to diurnal cycles in O_2 and CO_2 (Setter *et al.*, 1987). During the periods the plants produce or consume the gases there will be concentration gradient between the environment and the plant tissues.

Waterlogging results in many changes in the soil (Jackson and Drew, 1984) (Figure 1.1) such as:

1. Decrease of (O_2) consumed by the plant roots and micro-organisms;
2. Increase of ethylene (C_2H_4): a gaseous plant hormone;
3. Increase of (CO_2) produced by the plants and micro-organisms.

This figure does not include the increased susceptibility of the tissues to pathogens (Drew and Lynch, 1980).

The decrease in O_2 results in other changes in the soil, such as a reduction in redox potential (*i.e.* to more negative values) and production of toxic end products of anaerobic metabolism by bacteria (Drew, 1983). Plants also become often more susceptible to microbial attack (Drew and Lynch 1980). The reduction in redox potential is caused mainly by micro-organisms, which can use compounds like NO_3^- and Fe_3^+ instead of O_2 as electron acceptors in the electron transport chain (Ponnamperuma, 1984). These chances can have profound effects on plant health.

In agricultural soils transient waterlogging exerts its adverse effects on the plants mainly due to direct effects of low O_2 on the metabolism of the roots and to a lesser extent due to high concentrations of ethylene. Typical symptoms are reduced growth of roots, chlorosis of leaves and reduced leaf expansion (Drew, 1983, Jackson and

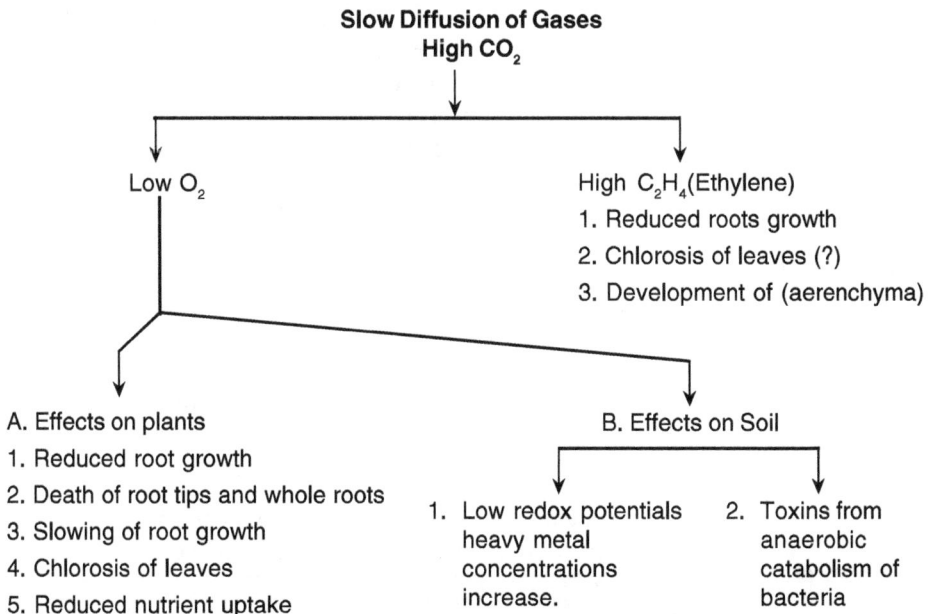

Slow Diffusion of Gases
High CO$_2$

Low O$_2$

High C$_2$H$_4$(Ethylene)
1. Reduced roots growth
2. Chlorosis of leaves (?)
3. Development of (aerenchyma)

A. Effects on plants
1. Reduced root growth
2. Death of root tips and whole roots
3. Slowing of root growth
4. Chlorosis of leaves
5. Reduced nutrient uptake

B. Effects on Soil

1. Low redox potentials heavy metal concentrations increase.

2. Toxins from anaerobic catabolism of bacteria

Figure 1.1: Summary of Effects of Waterlogging on Plants and Soils

Drew, 1983). These latter reductions may be due to reduced mineral nutrition, alternatively, there is feedback control to prevent a severe imbalance of shoot and root growth.

Ethylene also induces several responses conductive to adaptation to waterlogging, these include formation of aerenchyma and probably formation of adventitious roots. During long-term waterlogging, the decrease in redox potential and the production of toxins become, probably, more dominent than mere O_2 deficiency of the plant roots. It is evident therefore from the available published information that one of the effects of submergence of the foliage is to aggravate effects of waterlogging soils, by reducing ventillation and O_2 supply to the roots during dark period.

Roots activity and porosity under waterlogged situation become very critical and the plant intolerance to tolerance/resistance to water submergence are more linked to these root characteristics. Hence a simple method of monitoring root activity and porosity may go a long way in fast screening of genotypes for water submergence resistance. Srivastava (1995) has developed a method and formula for determining these root activity and root porosity in plants which could help screening genotypes for water submergence tolerance as described under:

Root Activity

The quantitative assessment of the status of reduction product, formazen giving pinkish colour with 2-3,5 triphenyl tetrazolium chloride (TTC) as a measure of root dehydrogenase activity in freshly detached roots incubated at $20\pm1°C$ has been used as an index of root activity.

The 3 mm apical tip of each root is excised, kept in cool chamber for 2-5 mts and plunged into 1 ml 0.2 per cent aqueous solution of 2-3,5 triphenyl tetrazolium chloride (TTC) and incubated at $20\pm1°C$ temperature in an incubator. The pinkish coloured formazen extracted from the root tips in 5 ml methyl cellosolve [2-mehoxy ethanol (absolute)] and the colour intensity which is directly proportional to dehydrogenase activity is read at 480 nm wave length on spectrophotometer. For numerical quantification it may be taken as, on unit equal to the change in optical density of 0.001 (Srivastava, 1995).

Root Porosity

Sundried root samples, devoid of any soil/sand particles (dried for 12 hours) are packed in 20 ml graduated test tube, 10 ml of luke warm distilled water (water temperature 35°C) is poured in the test tube and finally stacked in the water bath maintained at a temperature of $35\pm-1°C$. After 2 mts, the final volume is measured and the same set is left in the water bath for 60 mts. before the final volume is recorded. Root porosity is calculated using the formula, as described by (Srivastava, 1995).

$$\text{Root Porosity} = \frac{\begin{array}{c}\text{Volume of water two mts after}\\\text{root submergence}\end{array} - \begin{array}{c}\text{Initial volume of water}\\\text{poured in the test tube}\end{array}}{\begin{array}{c}\text{Volume of water two mts after}\\\text{root submergence}\end{array} - \begin{array}{c}\text{Volume of water 60 mts.}\\\text{after root submergence}\end{array}}$$

With series of studies it has been evident that one of the major biological consequences of flooding is tissue anoxia. Anoxia interferes with respiration at the level of electron transport (Albert *et al.*, 1983) and oxidative phosphorylation shalted, leading to reduction in energy charges (Pradet, 1978, Van Toai, 1989). All stages of the oxidative chain become saturated with electrons, resulting in the accumulation of NADPH (Crawford, 1985) and NADH (Albert *et al.*, 1983). Intercellular pH decreases (Robert *et al.*, 1985).

Probably as result of the accumulation of lactate (Hoffman *et al.*, 1986) and free protons associated with NADH, most plant tissues under anoxia shift to alcoholic fermentation, probably to maintain substrate level phosphorylation and redox cycling (Davies *et al.*, 1985). Oxygen toxicity has been proposed as the principal mechanism of post anoxic injury. Superoxide radical (O_2^-) the first intermediate in the reduction of diatomic oxygen, is highly reactive and acts as a precursor of other toxic intermediates (Elstner, 1987). Superoxide radicals are generated when xanthine dehydrogenase is proteolytically cleaved under conditions of low energy charge (Roy and Mc Cord, 1978). In addition, electron transport carriers, when saturated, undergo autooxidation and transfer, reducing equivalents to oxygen to produce O_2^-. Thus the conditions favourable to O_2^- formation are high reducing equivalent levels, low energy charge and saturated electron transport components–all exists, in anoxic plant tissues. When aerobic conditions are restored from an anoxic stage, a burst in O_2^- production occurs, resulting in post anoxic injury to the tissues (Babior, 1987).

Living cells have evolved mechanism for protection against oxygen toxicity. The first enzyme is the free radical scavenging pathway superoxide (SOD) which is actually a group of multimeric metalloenzymes that catalyse the reaction $2O_2^- + 2H^+ - H_2O + O_2$. SOD is inducible by hyperbaric oxygen conditions (Fridovich, 1986) or by treatment of the cells with dinitrophenol which uncouples the electron transport chain (Dryer *et al.*, 1990).

The increase in SOD activity under these conditions ae presumed to protect the cells against the deleterious effects of O_2^- produced upon their return to air.

In reviewing mechanisms of anoxic tolerance, the universal requirements for retention of membrane integrity is an important issue despite a greatly reduced energy supply and reductions in the requirements of energy for maintenance. Using these requirements as a pre-requisite, there appears to be two different modes of adaptation to anoxic, based on (*i*) rapid and (*ii*) slow rates of catabolism respectively. Anoxia tolerant plants with the mode of slow rates of catabolism may be particularly useful during transient O_2 deficiency as often occurs in flood prone rainfed environment.

Injury due to exposure to anoxia can also be aggravated upon return from anoxia to aerated conditions, for example when production of free radicals of oxygen would aggravate membrane damage and injury in several species. Specifically, partial break down of fatty acids of membranes during anoxia may lead to formation of free radicals of oxygen leading to further damage to membranes. Strong evidence of this possibility has been reported by Monkel *et al.*, 1987, Crawford, 1993) Crawford and Wollenweber, Ratzer, 1992 and Monk *et al.*, 1989). The response to anoxin, which involves a large reduction in energy production is, also relevant to other situations, for example to carbohydrate starvation, which can in principle occur due to other factors of a submerged environment such as low CO_2, supply and low light.

Treatment with ascorbic acid(AA), chick pea seedlings before re-exposure to air after varying periods of anoxia, shows that this antioxidant can improve seedling growth and survival during post anoxia recovery period (Crawford and Wollenweber-Ratzer, 1992). There is dearrangement in hormonal metabolism of the plants ground under flooding situation (Phillips, 1964; Crozier *et al.*, 1969). Attempts have been made to alleviate the deleterious effects of waterlogging by applying hormones (Jackson and Compbell, 1979). There is an increase in stomatal conductance which may develop due to the accumulation of abscisic acid or reduction in the levels of gibberellins and cytokinine as observed in waterlogged plants (Reid and Bradford, 1984). Application of CCC(Cycocoel), a growth retardant may enhance gibberellin content of the plant body (Van Dragt, 1969), which may help us retaining stomatal conductance.

Identification of the problem has enabled plant physiologist to try and target limiting metabolic processes in different genotypes and develop physiological tools for screening.The last adaptation to anaerobiosis in response to water submergence stress as presumed has a specific relationship with root activity and root porosity in a wide ranging variability in genotypes and their response to water submergence (Srivastava, 1995). The changing level of carbohydrates in the genotypes as an index for submergence tolerance can also rapidly be measured using Near Infrared Reflectance Spectroscopy.

Achievements made in understanding the physiology and biochemistry of submergence tolerance and tracing the energy supply pathway through alcoholic fermentation during anoxia has developed linkages for further research in enzymology to pinpoint important metabolic steps which limits metabolism in plants under anoxia which would go a long way in linking this to molecular biology for cloning and transforming greater number of copies of the appropriate genes for submergence tolerance.

The expected outcome of this research is to locate the gene for submergence tolerance on the crop genome map and to "tag" it with restriction fragment length polymorphisms (RFLPs). Once linkage has been established between a DNA fragment and the phenotype, the probe can be used in later breeding programs for marker-assisted selection. Physiology studies have demonstrated the importance of alcoholic fermentation (AF) during submergence and its linkage in the process gives an indication that one of the genes encoding the enzymes in AF, in more particular the pyruvate decarboxylase (PDC) over alcoholic dehydrogenase (ADH) may prove to be responsible for submergence tolerance.

An effort in molecular biology approaches to submergence tolerance is therefore warranted since the production of a submergence tolerant genotype via-plant breeding may necessiate a considerable effort to rebuild the genotype for agronomic fitness. However, it is clear that the transgenic plants with altered PDC/ADH levels will require careful, physiological evaluation to relate with the level of submergence tolerance produced. It is reported in maize that 10 or more major anaerobic genes are triggered during anoxia but the concerned induction of anaerobic ally induced genes, and the observation that several of these genes show regions of sequence similarly in their promoters are consistant with the hypothesis that a common transcription factor is involved. Differences in the efficiency of such a factor could account for differences in submergence tolerance. Hence, biotechnology can prove to be a useful tool to delineate the factors and help genetic manipulations to increase submergence tolerance in crops producing specific genetic mutations using transgenic plants by RFL-type of techniques to identify the key gene(s) for submergence tolerance.

References

Albert, B., Bray, D., Lewis, J., Raff, M., Roberts, K. and Watson, J., 1983. *Molecular Biology of the Cells*. Garland Publishing Inc., New York, pp 67–80.

Babior, B.M., 1987. The respiratory burst oxidase. *Trends Biochem. Sci.*, 12: 241–243.

Crawford, R.M.M., 1985. Effects of environmental stress on lipid metabolism in higher plants. *Agrochimica*, 2: 51–63.

Crawford, R.M.M. and Wollenweber-Ratzer, 1992. Influence of L-ascrobic acid on post anoxic growth and survival of chickpea seedlings (*Cicer arietinum*). *J. Exp. Bot.*, 43: 703–708.

Crozier, A., Reid, D.M. and Harvey, B.M., 1969. Effects of flooding on the export of gibberellins from the root to the shoot. *Planta*, 89: 376.

Davies, D.D., Kenworthy, P., Mocquot, B., Roberts, K., 1985. The metabolism of pea roots under hypoxia. In: *Current Topics in Plant Biochemistry and Physiology*, Vol. 4, (Eds.) D.D. Randall, D.G. Belvins and W.H Campbell. University of Missouri, Columbia, MO, pp. 141–155.

Drew, M.C., 1983. Plant injury and adaptation to oxygen in the root environment: A review. *Plant and Soil*, 75: 179–199.

Drew, M.C. and Lynch, J.M., 1980. Soil anaerobiosis, microorganisms and root function. *Ann. Rev. Phytopathology*, 18: 37–66.

Dryer, S.E., Dryer, R.L. and Autor, A.P., 1980. Enhancement of mitocondrial cyanide-resistant superoxide dismutase in the liver of rats treated with 2,4-dinitrophenol. *J. Biol. Chem.*, 255: 1054–1057.

Elstner, E.F., 1987. Metabolism of activated oxygen species. In: *The Biochemistry of Plants*, Vol. 11. Academic Press Inc., New York, pp. 253–314.

Fridovich, I., 1986. Superoxide dismutases. *Adv. Enzymol.*, 51: 61–97.

Hoffman, N.E., Bent, A.F. and Hanson, A.D., 1986. Induction of lactase dehydrogenase isozymes by oxygen deficit in barley root tissues. *Plant Physiol.*, 82: 658–663.

Jackson, M.D. and Campbell, D.I., 1979. Effect of Bezyladenine and Gibberellic acid on the response of plants to anaerobic environment and to ethylene. *New Phytol.*, 82: 331–340.

Jackson, M.B. and Drew, M.C., 1984. Effects of flooding on growth and metabolism of herbaceous plants. In: *Flooding and Plant Growth*, (Ed.) T.T. Kozlowaski. Academic Press.

Keeley, J.E. and Franz, E.H., 1979. Alcoholic fermentation in swamp and upland populations of *Nyssa sylvafica*: Temporal changes in adaptive strategy. *Amer. Nat.*, 113: 587–592.

Kozlowaski, T.T., 1984. In: *Flooding and Plant Growth*, (Ed.) T.T. Kozlowaski. Academic Press.

Lie, T.A., 1984. Biological nitrogen fixation. In: *Crop Physiology*, (Ed.) U.S. Gupta, pp. 97–131.

Ponnamperuma, F.N., 1984. Effect of flooding on soils. In: *Flooding and Plant Growth*, (Ed.) T.T. Kozlowaski. Academic Press.

Pradet, A. and Bomsel, J.L., 1978. Energy metabolism in plants under hypoxia and anoxia. In: (Eds.) *Plant Life in Anaerobic Environments*, D.D. Hook and R.M.M. Crawford. Ann Arbor Science Publ., Ann Arbor, MI, pp. 89–118.

Reid, D.M. and Bradford, K.J., 1984. Effects of flooding on hormone relations. In: *Flooding and Plant Growth*, (Ed.) T.T. Kozlowaski. Academic Press, London, pp. 195–219.

Roberts, J.K.M., Andrade, F.H. and Anderson, I.C., 1985. Further evidence that cytoplasmic acidosis is a determinant of flooding intolerance in plants. *Plant Physiol.*, 77: 492–494.

Roy, R.S., and MeCard, J.M., 1978. Ischemia-induced conversion of zanthine dehydrogenase to zanthine oxidase. *Fed. Proc.*, 41: 767–772.

Setter, T.L., Kupkanchanakul, T., Kupkanchanakul, K., Bhekasut, P., Wiengweera, A. and Creenway, H., 1987. Concentrations of CO_2 and O_2 in floodwater and in Internodal lacunae of floating rice growing at 1-2 metre water depths. *Plant Cell and Environment*, 10: 767–776.

Srivastava, Ashwani K., 1995. Method for measuring root activity and root porosity in plants subjected to water submergence stress (Unpublished).

Van Toai, T.T., 1989. Extraction and determination of seed adenine nucleotides by different methods for anaerobic stress evaluation. *Seed Sci. Technol.*, 17: 439–451.

Chapter 2

Reclamation and Management of Saline and Alkali Soils

S.B. Agrawal and Anoop Singh

Department of Botany,
Allahabad Agricultural Institute (Deemed University),
Allahabad – 211 007

ABSTRACT

Soil is the most precious natural resource and thus requires proper management. Estimates show that the world as a whole is losing at least 3.0 ha of arable land every minute due to salinization or sodification. In India about 7.0 M ha land is affected by salinity and alkalinity. The problem of saline and alkali soils is old but its magnitude and intensity have been increasing because of poor land and water management practices. The proper land management by way of its reclamation involves physical, chemical and biological means, which are site specific and their integration is highly desirable to combat with the problem. The present review is an attempt to emphasize the problem of salinity and alkalinity of soils, its effect on plants and application of physical, chemical and biological methods of soil reclamation along with management issues.

Keywords: *Reclamation, Saline soil, Alkali soil, Management, Constraints.*

Introduction

The formation of one-inch soil takes about 400 years but it degrades in few years only. As a result of our ever-increasing population, the demand for food is increasing day-by-day, for this we have two options either to increase the land area or to produce more food per unit area. The later option can be exercised but only up to a certain extent. As for the first option the land area is already finite. National waste lands development board estimated that about 175 M ha out of 329 M ha of total land of India belonged to the category of waste lands which yielded zero to a very nominal level, out of which 7 M ha is saline or alkali, and the largest area of about 1.25 M ha is found in U.P., which comes about 4 percent of its total geographical land. So we can increase the production by reclaiming such problematic soils. This can be done by various land reclamation practices some are traditional and some modern and it is equally important to judiciously manage the soil after reclamation.

Soils that contain excess soluble salts and affect plant growth adversely are called salt affected soils. These soils are mainly divided into two major categories.

Saline Soil

Saline soils are those which contain sufficient soluble salts (chlorides and sulphates of sodium) to interfere with the growth of plants. These soils are locally called as *Reh*. The six classes of saline soils found in India are (1) saline soils (illitic) of semiarid regions with highly saline ground water (2) saline soils (illitic) of sub humid region with moderately saline groundwater (3) saline soils (illitic) of sub humid region (4) saline soils (montmorillonitic) of semi arid delta region (5) saline acid sulfate soils of humid region and (6) saline marsh of arid region.

Alkali Soils

These are also called as sodic soils and these soils contain excessive exchangeable sodium to adversely affect crop production and locally called as *User*. The alkali soils found in India fall under three classes, *viz.* (1) alkali soils (illitic) of Indo Gangatic plain with sweet ground water, (2) alkali soils (illitic) of Indo Gangatic plain with sodic ground water and (3) alkali soils (montmorillonitic) of deccan plateaus with sodic or saline ground water.

Balba (1995) compared the saline and alkali soils with normal soil and conclude that pH and EC of saline soil is more than the normal soil due to presence of more sodium salts and alkali soils have more pH and SAR than the normal soils due to presence of excessive exchangeable sodium salts in the soil (Table 2.1). Properties of these problematic soils are summarized in Table 2.2.

**Table 2.1: Properties of Normal Soils Compared with Saline and Alkali Soils
(Source: Balba, 1995)**

Soil	pH	EC (dSM^{-1})	SAR
Normal	6.5	<4	<15
Saline	7.3-8.5	>4	<15
Alkali	>8.5	<4	>15

Table 2.2: Properties of Saline and Alkali Soils
(*Source*: Abrol, 1998)

Sl.No.	Properties	Saline Soil	Alkali Soil
1.	Chemistry of soil solution	Dominated by chlorides and sulphides of sodium, calcium and magnesium	Dominated by sodium carbonate or sodium bicarbonate or both
2.	Main adverse effect on plants	High osmotic pressure of soil solution	Alkalinity of soil solution and its corrosive effect on plant roots
3.	Aim of reclamation	Removal of excessive electrolytes through leaching	Lowering or neutralizing high pH through chemical amelioration

Factors Responsible for Formation of Saline and Alkali Soils

1. *Arid and semiarid climate*: In these climatic regions annual evaporation is more than the annual rainfall, resulting which salts present in the lower strata comes on the surface with evaporating water and accumulated on the surface.

2. *Poor drainage of soil*: Due to evaporation of stagnated water in poor drainage areas salts are accumulated on the surface of soil.

3. *High water table*: In high water table areas submerged conditions prevails after evaporation of water such soil becomes saline.

4. *Over flow of seawater over lands*: Seawater is brackish in nature and when it over flow on land, which ultimately increase the concentration of salts.

5. *Bad quality of irrigation water*: Soils become saline when farmers use bad quality water in irrigation, which contains salts in higher concentrations.

6. *Salts blown by wind*: In highly arid region those like Rajasthan, salts are present on the soil surface and blown by high velocity wind, which are deposited on another place.

7. *Saline nature of parent rock material*: When parent rock is saline, soil formed from these rocks is naturally saline.

8. *Continuous use of basic fertilizer*: Continuous application of basic fertilizer increases salts in the field, and makes them saline.

9. *Low permeability of soil to water*: Continuous ploughing at a certain depth forms a hard layer, which restricts the downward movement of water due to which salts are accumulated on surface by evaporation.

10. *Undulated land*: In undulated region water containing salts flows down from upper areas to lower areas, which increases more salts in lower areas.

11. *Presence of hard kanker pan in lower surface*: Due to presence of hard kanker pan in lower surface downward movement of water is inhibited resulting which salts are accumulated on surface by evaporation.

Sources of Salts in Soil

According to Balba (1995) sources of salts are mainly divided in to primary and secondry, primary sources includes parent rocks, marine source and deltaic source, while secondary sources include deforestation, overgrazing, disposal of brakish water, irrigation with salty water, inadequate drainage, etc. Srivastava and Srivastava (1993) reported that organic carbon decreases with increasing soil pH. Organic carbon and nitrogen have a positive correlation that's why as soil pH decreases nitrogen also decreases (Figure 2.1), resulting which soil become less fertile or unfertile.

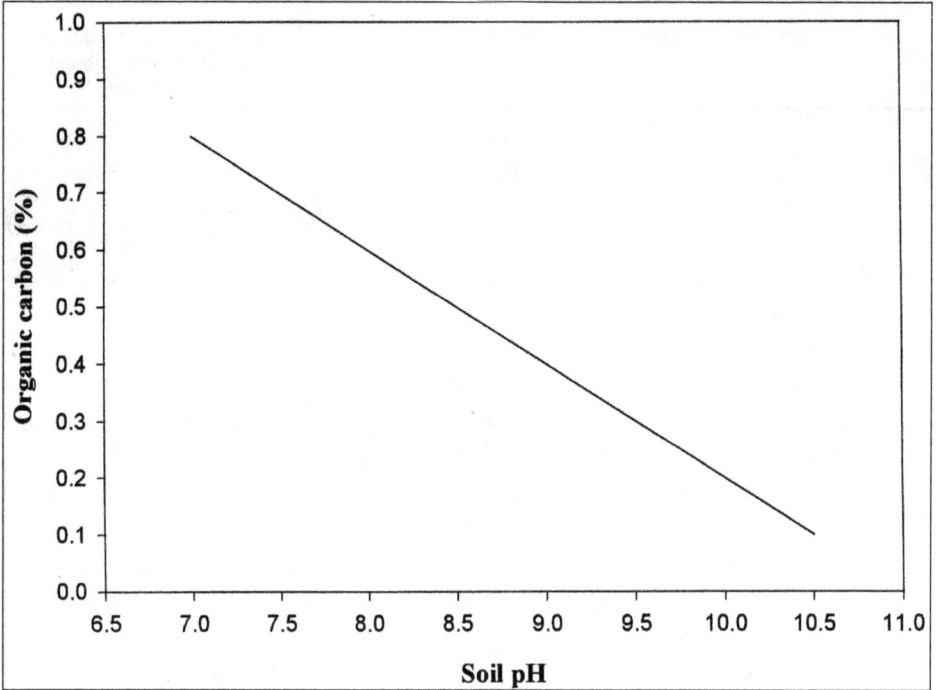

Figure 2.1: Effect of pH on Soil Organic Carbon
(*Source*: Srivastava and Srivastava, 1993)

Effects of Salts on Soil

When salts are accumulated in the soil (surface and subsurface), they reduces the availability of nutrients to plants, as they binded with them. Salts also decreases the activity of microorganisms. Salts are accumulated in subsurface soil and form hard pan, which restrict water to percolate at lower strata. As salts enriches in the soil, the physical properties (texture, porocity, bulk density, etc.) and chemical properties (pII, EC, SAR, etc.) also adversely affected.

Effects of Salts on Plants

As salts are enriched in soil, the osmotic pressure of soil water increases, and when this osmotic pressure is more than the osmotic pressure of roots, then plant is unable to absorb nutrients. Germination of seeds also reduces in salt affected soil because as radicle and plumule emerges, they are injured with salts. Availability of some elements (Ca, Mg etc.) decreases and few elements (Na etc.) increases up to toxic levels. Some saline and alkali hazards on plants are given in Table 2.3.

Table 2.3: Saline and Alkali Hazards on Plants in Relation to ESP and EC
(*Source*: Balba, 1995)

ESP	EC(dSM⁻¹)	Saline/Alkali Hazard
< 15	<2	None to slight
15-30	2-4	Slight to moderate
30-50	4-8	Moderate to high
50-70	8-16	High to very high
70	> 16	Extremely high

Srivastava and Srivastava (1993) reported that organic carbon decreases with increasing soil pH. Organic carbon and nitrogen have a positive correlation that's why as soil pH decreases nitrogen is also decreases (Figure 2.1), resulting which soil become less fertile or unfertile. Chaney *et al.* (1992) reported that as HCO_3 increases in the soil, soil pH increases and create buffer, which reduces the availability of Fe III and roots secret phenolic substances, resulting which synthesis of CO_2 and organic acid increases and whatever amount of Fe III is reached in roots a part of that amalgamated with organic acid and rest go to shoot this Fe III is uneven distributed in leaves that's why chlorophyll is formed in patches, so yield of plant is decreased (Figure 2.2).

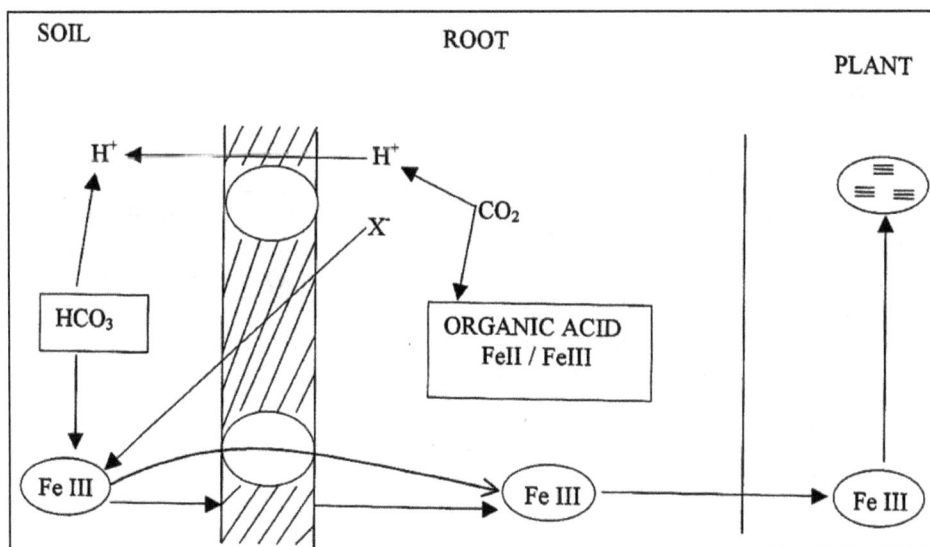

Figure 2.2: Effects of Bicarbonate on Chlorophyll Formation in Plants
(*Source*: Chaney *et al.*, 1992)

Methods of Reclamation

Reclamation methods are mainly divided into three *viz.* Physical, chemical and biological. Physical methods are crude and time consuming but these are most suitable for the reclamation of saline soils while chemical methods give quick response and suitable for reclamation of alkali soils. Biological methods are ecofriendly and reclaim both type of soils but they are time consuming.

Physical Methods

The following methods are found suitable to reclaim these problematic soils.

Scraping of salts

We scrape the salts and dump them elsewhere.

Leaching of the Salts

In this method field filled with water and salts are leached with water.

Drainage

In drainage huge amount of water leave in the field for a short period after that drained out in which salts are dissolved.

Trenching

We dig a trench and when we dig another trench the soil of this trench is filled in the first trench resulting which the lower soil becomes upper soil and vice versa.

Breaking of Hard Kanker Pan

The down ward movement of water along with salts is allowed by the breaking of hard kanker pan.

Deep Ploughing

Deep ploughing breaks subsurface layer by which downward movement of water along with salts is possible.

Chemical Methods

Optimum dose: The optimum doses of chemicals are depending upon three factors (*i*) Nature and degree of soil deterioration, (*ii*) Soil texture and (*iii*) Type of crop to be grown. Each chemical firstly formed calcium sulphate ($CaSO_4$), then this $CaSO_4$ react with sodium binded clay particles and replaced Na^+ with Ca^{++} and formed sodium sulphate (Na_2SO_4), which percolate in lower stratas with water and/or drained out from the field through facilitating drainage.

Use of Gypsum

$$2NaHCO_3 + CaSO_4 \longrightarrow CaCO_3 + Na_2SO_4 + CO_2$$

$$Na_2CO_3 + CaSO_4 \longrightarrow CaCO_3 + Na_2SO_4$$

Na^+

$$\boxed{CLAY} + CaSO_4 \longrightarrow Ca^{++}\boxed{CLAY} + 2Na_2SO_4$$
Na^+

Use of Pyrite

$$FeS_2 + 2H_2O + 7O_2 \longrightarrow FeSO_4 + 2H_2SO_4$$

$$\begin{matrix} Na^+ \\ \boxed{CLAY} \\ Na^+ \end{matrix} + CaSO_4 \longrightarrow Ca^{++}\boxed{CLAY} + 2Na_2SO_4$$

Use of Sulphur

$$2S + 3O_2 \longrightarrow SO_3$$

$$SO_3 + H_2O \longrightarrow H_2SO_4$$

$$CaCO_3 + H_2SO_4 \longrightarrow CaSO_4 + CO_2 + H_2O$$

$$\begin{matrix} Na^+ \\ \boxed{CLAY} \\ Na^+ \end{matrix} + CaSO_4 \longrightarrow Ca^{++}\boxed{CLAY} + 2Na_2SO_4$$

Use of Sulphuric Acid

$$CaCO_3 + H_2SO_4 \longrightarrow CaSO_4 + CO_2 + H_2O$$

$$\begin{matrix} Na^+ \\ \boxed{CLAY} \\ Na^+ \end{matrix} + CaSO_4 \longrightarrow Ca^{++}\boxed{CLAY} + 2Na_2SO_4$$

Use of Calcium Polysulphide

$$CaS_5 + SO_2 + 4H_2O \longrightarrow CaSO_4 + H_2SO_4$$

$$CaCO_3 + H_2SO_4 \longrightarrow CaSO_4 + CO_2 + H_2O$$

$$\begin{matrix} Na^+ \\ \boxed{CLAY} \\ Na^+ \end{matrix} + CaSO_4 \longrightarrow Ca^{++}\boxed{CLAY} + 2Na_2SO_4$$

Use of Ferrous Sulphate

$$FeSO_4 + H_2O \longrightarrow FeO + H_2SO_4$$

$$CaCO_3 + H_2SO_4 \longrightarrow CaSO_4 + CO_2 + H_2O$$

$$\begin{matrix} Na^+ \\ \boxed{CLAY} \\ Na^+ \end{matrix} + CaSO_4 \longrightarrow Ca^{++}\boxed{CLAY} + 2Na_2SO_4$$

Abrol *et al.* (1998) studied the relationship between chemical requirement with soil pH and texture and reported that chemical requirement increases with increasing pH and more chemicals require for reclamation of heavy soils followed by medium

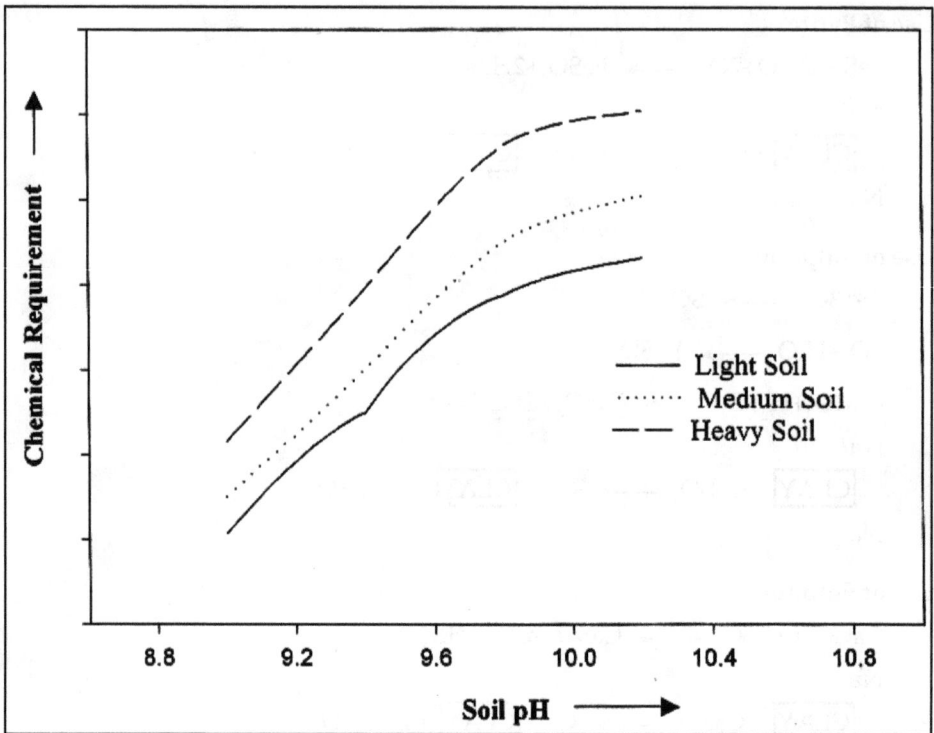

**Figure 2.3: Relationship Between Chemical Requirement with Soil pH and Texture
(Source: Abrol *et al.*, 1998)**

and light soil (Figure 2.3). Math *et al.* (1999) conducted a experiment on gypsum application and concluded that gypsum application decreases pH, EC and ESP of soil (Table 2.4)

**Table 2.4: Effect of Gypsum Application on Soil Properties
(*Source*: Math *et al.*, 1999)**

pH		EC (dSM⁻¹)		ESP	
EGA	AGA	BGA	AGA	BGA	AGA
8.7	8.1	1.21	1.00	20.0	16.0
8.5	8.0	0.50	0.30	31.0	24.0
9.2	8.5	2.40	0.86	15.8	10.5
9.5	8.9	1.90	0.35	19.9	14.4
9.6	8.8	2.40	0.84	28.4	14.5
9.1	8.5	1.68	0.67	23.0	15.9

BGA: Before gypsum application.

AGA: After gypsum application.

Abrol *et al.* (1998) reported that by the application of FYM in combination of gypsum, the growth of *Casuarina equisetifoliya* plant is maximum with 100 per cent survibility, while by the application of gypsum alone survibility is 100 per cent but growth is less and by the application of FYM alone and in control survibility is 82 per cent and 60 per cent, respectively (Figure 2.4).

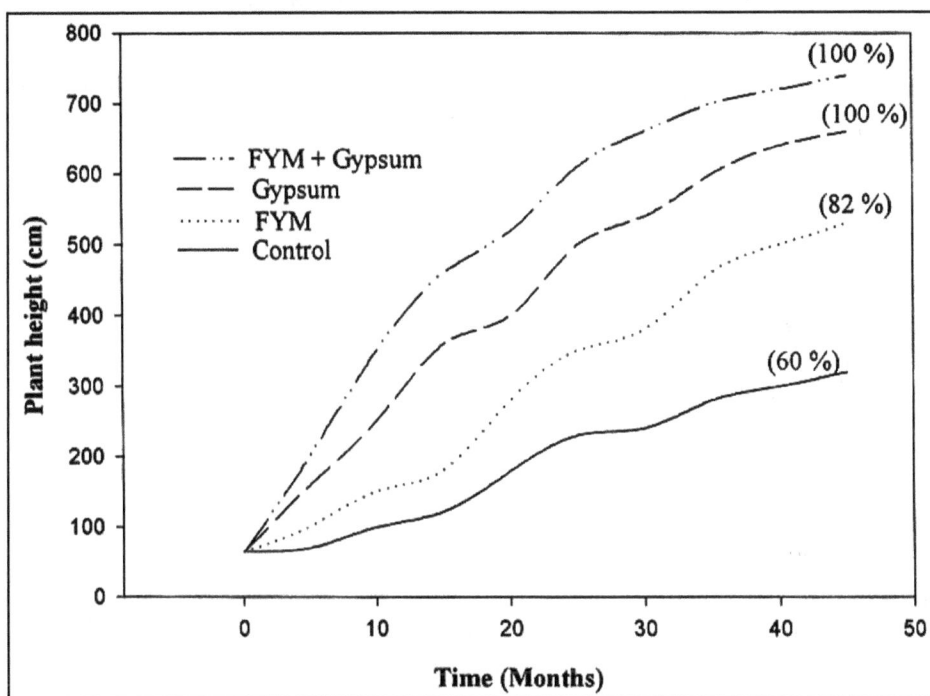

**Figure 2.4: Effect of Gypsum and FYM Application on the Growth of
Casuarina equiselifolia in Sodic Soils
(Values in parenthesis showing germination percentage)
(*Source*: Abrol *et al.*, 1998)**

Biological Methods

Use of Organic Manure

After decomposition of organic manure like FYM compost etc humic and fulvic acids are formed which reclaims soil.

Use of Green Manure

By the use of green manure, chemical and physical properties of soil are improved and also increase microbial activity.

Use of Molasses

By the addition of molasses organic acids are formed which reclaims the soil.

Mulching

Mulching directly reduces evaporation rate and also adds organic matter after decomposition.

Use of Crop Residue

Soil can also be reclaimed by use of crop residues directly or in the form of mulching, organic manure etc.

Growing of Weeds

The roots of weeds like *Argemone maxicana* secrets chemicals, which reclaim soil.

Growing of Salt Tolerant Plants

Soil can also be reclaimed by growing of salt tolerant plants but it takes long time.

Tolerant Plants

Trees

1.	Babool	2.	Jhau
3.	Black- Siris	4.	Arjun

Grasses and Forage Crops

1.	Alfa-alfa	2.	Karnal grass
3.	Bermuda grass	4.	Para grass
5.	Napier grass	6.	Clover

Cereal Crops

1.	Barley	2.	Rice
3.	Sorghum	4.	Wheat

Vegetable Crops

1.	Carrot	2.	Lettuce
3.	Potato	4.	Onion

Fruit Crops

1.	Ber	2.	Aonla
3.	Guava	4.	Phalsa
5.	Karonda		

Table 2.5: Effect of Mulching on Evaporation Rate and Soil EC
(*Source*: Kanthaliya *et al.*, 1990)

Depth of Water	Evaporation Rate mm/day		$EC(dSM^{-1})$	
Table (cm)	UMS	MS	UMS	MS
25	4.0	1.1	10.9	9.3
50	4.0	1.4	10.0	9.0
75	2.9	1.3	7.0	5.1
100	1.7	1.1	5.2	2.7

UMS: Unmulched soil.

MS: Mulched soil.

Kanthaliya *et al.* (1990) studied the effect of mulching at different depth of water table and reported that mulching reduces evaporation rate and EC of soil at each depth of water table (Table 2.5). Mongia and Chhabra (2000) analyzed the effect of agroforestry on chemical composition of alkali soil profile and found that by agroforestry pH and EC of soil reduces at each depth of soil profile (Table 2.6). Khanna (1994) conducted a experiment with various fruit crops model in sodic soils and after five year he reported that each fruit crops model reduces the pH, EC and ESP of soil at each soil depth but Guava and Ber model gave best result among them (Table 2.7).

Table 2.6: Effect of Reclamation on Chemical Composition of Alkali Soil Profile through Agroforestry (*Source*: Mongia and Chhabra, 2000)

Depth of Soil (cm)	pH		EC (dSM⁻¹)	
	BFA	AAF	BAF	AAF
0–15	10.6	8.7	13.2	7.5
15–30	10.5	8.5	6.3	3.8
30–45	10.5	8.4	4.1	3.2
45–60	10.4	9.0	2.9	2.0
60–90	10.4	9.4	2.2	1.2
90–120	10.3	9.5	1.9	1.0
120–150	10.3	9.5	1.6	0.9
150–180	10.2	9.4	1.6	0.7

BAF: Before agroforestry.

AAF: After agroforestry.

Table 2.7: Change in Properties of Sodic Soil Under Various Fruit Crop Model After Five Year (*Source*: Khanna, 1994)

Model	Soil Depth (cm)	pH	EC (dSM⁻¹)	ESP
Control	0–15	10.4	18.3	76
	15–30	10.7	21.3	82
	30–45	10.6	24.2	84
Aonla + Guava	0–15	9.2	13.4	60
	15–30	10.0	15.5	66
	30–45	10.3	18.4	75
Aonla + Ber	0–15	9.5	12.0	57
	15–30	9.8	13.0	65
	30–45	10.2	15.0	73
Guava + Ber	0–15	9.0	8.3	47
	15–30	9.2	10.5	56
	30–45	9.6	14.0	65

Management

Proper management is essential after reclamation of such problematic soils therefore by adopting following techniques, soils can be managed successfully.

Ploughing

Ploughing should be avoided at same depth.

Puddling

Excessive puddling should be avoided.

Light and Frequent Irrigation

Irrigation should be applied lightly at frequent interval, which reduces evaporation.

Regular Cultivation

Regular foliage cover achieved by adopting intensive crop rotation, so reducing the rate of evaporation.

Method of Irrigation

It is beneficial to provide minimum water for irrigating the crops, for this drip and sprinkler irrigation method gave best result.

Drainage

Regular drainage is essential to avoid stagnation of excessive water in the field.

Fertilizers

Use of basic fertilizers like $CaCO_3$, basic slag etc should not be applied continuously.

Balba (1995) conducted a experiment using different bed shapes along with irrigation methods and reported that by using sloping bed, plants can be grown up to maximum salinity level *i.e.* 16 dSM^{-1} white by using rectangular bed, plants can be grown up to 8 dSM^{-1} in double row system and only up to 4 dSM^{-1} in single row system irrespective of irrigation methods (Figure 2.5).

Some Constraints Faced by Farmers in Adoption of Technology

1. The economy used by farmers in the use of fertilizers, amendment and land leveling is not cost effective.
2. Difficulty and inhibition on the part of farmers to get credit.
3. Fears that after reclamation, land would attract the provision of land ceiling act.
4. Long time taken by concerned department to supply electric connection to run tube well.
5. Lack of specialized extension staff earmarked to take up land reclamation in the state department of agriculture.

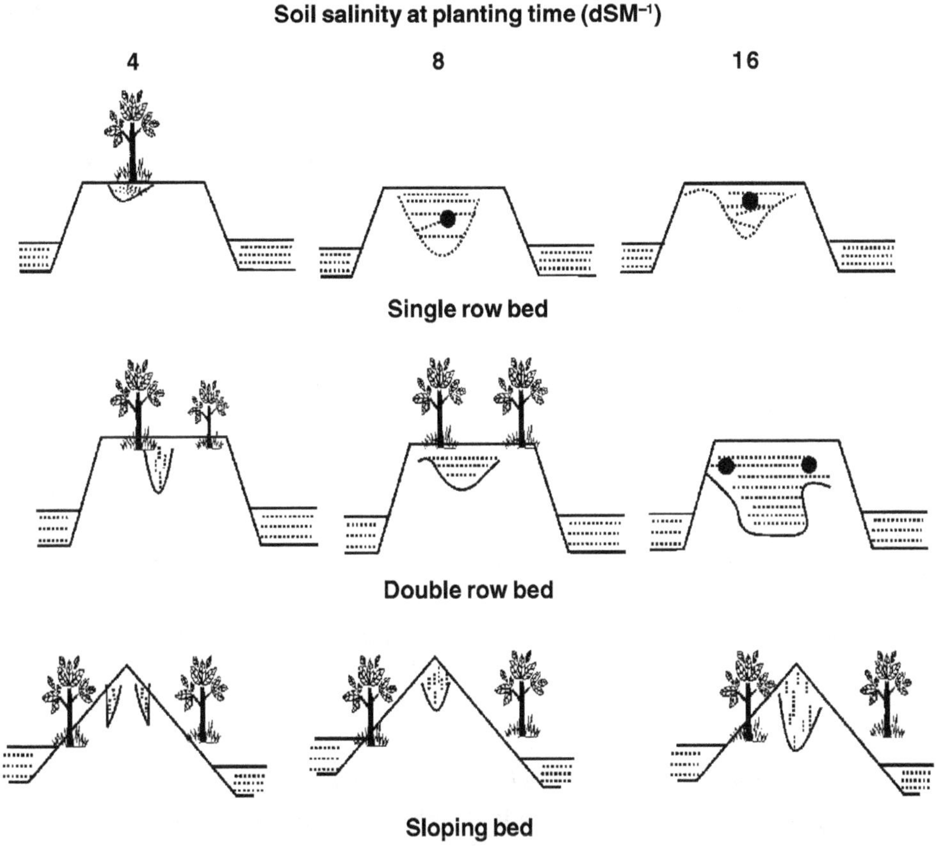

Figure 2.5: Bed Shapes and Salinity Effects
(*Source*: Balba, 1995)

Conclusion

The salt affected soils present different kinds and intensities of problems. They are inherently rich in soil fertility and are high in production potential. However, these soils produce practicably nothing at present except some little grass. Though these soils may not produce bumper crops in the first year of reclamation but there will be gradual increase in production. For this specific ameliorative efforts requires like horizontal flushing and reclamation of soil by using physical, chemical and biological methods in combination and also need to develop appropriate policies and implement them. The reclamation of such soils in five-year plan is essential for the more effective utilization and conservation of land and water resources. Given a chance to reclaim these lands will considerably aid to the production of the country.

References

Abrol, I.P., Gupta, R.K., Jhoshi, P.K. and Prasad, R., 1998. Salt affected soils and their management. In: *Technologies for Wasteland Development*, (Eds.) I.P. Abrol and V.V.D. Narayana. ICAR, New Delhi, p. 307–380.

Annonymous, 1989. *Making Usar Bloom.* U.P. Council of Agricultural Research. Published by Hindustani Book Depot, Lucknow.

Balba, A.M., 1995. Salt affected soils. In: *Management of Problem Soils in Arid Ecosystem.* Lewis Publishers, New York, p. 20–79.

Brady, N.C., 1996. Soil reaction: acidity and alkalinity. In: *The Nature and Properties of Soils,* 10th Edn. Prentice Hall of India, New Delhi.

Cantero, A., Cantero, J.J. and Cisneros, J.M., 1999. Vegetation, soil hydrophysical properties, and grazing relationship in saline-sodic soils of Central Argentina. *Canadian Journal of Soil Science,* 79(30): 399–409.

Chaney, R.L., Coulombe, B.A., Bell, P.F. and Angle, J.S., 1992. Detailed method of screen dicot cultivars for resistance to Fe chlorosis using Fe-EDTA and bicarbonate in nutrient solution. *Journal of Plant Nutrition,* 15: 2063–2083.

Kanthaliya, P.C., Chippa, B.R. and Sharma, S.L., 1990. Effect of water table depth and salinity of ground water on accumulation of salts in mulched and unmulched soils. *Indian Journal of Agricultural Chemistry,* 23(1): 45–50.

Khanna, S.S., 1994. Management of sodic soils through tree plantation. *Journal of the Indian Society of Soil Science,* 42(3): 498–508.

Math, S.K.N., Rao, M.S.M., Padmaiah, M. and Patil, S.L., 1999. Effect of gypsum on soil properties and yield of rice in the loamy soils of semiarid tropics of Andhra Pradesh. *Journal of Soil and Crops,* 9(2): 145–150.

Mongia, A.D. and Chhabra, R., 2000. Silica and phosphate profiles of alkali soils following reclamation. *Journal of the Indian Society of Soil Science,* 48(1): 33–37.

Qadir, M., Qureshi, R.H. and Ahmad, N., 1998. Horizontal flushing a Promising ameliorative technology for hard saline-sodic and sodic soils. *Soil and Tillage Research,* 45: 119–131.

Sinha, R.L., 1963. The fundamentals of conservation in saline and alkali soils. *Journal of Soil and Water Conservation in India,* 3(3–4): 70–75.

Srivastava, A.K. and Srivastava, O.P., 1993. Transformation and availability of nitrogen in salt affected soil. *Indian Journal of Agricultural Chemistry,* 26(2 and 3): 55–61.

Tripathi, B.R., 1998. Managing sodic soils for sustained crop production in Indo-Gangatic Plains. *Journal of the Indian Society of Soil Science,* 46(4): 543–550.

Chapter 3

Degradation of Reserved Forest (Victoria Park) Near Bhavnagar City Due to Overexploitation, Gujarat, India

Renu A. Oza

Department of Botany, Sir P.P Institute of Science,
Bhavnagar University, Bhavnagar – 364 002

ABSTRACT

The reserved forest [Victoria park] situated at distance of 3km, south of Bhavnagar city, Gujarat State, India. Bhavnagar lies on Western cost of India at 72°11' E Longitude and 21°45' N Latitude and 11M above sea level. It is semi arid region with very hot summer and cold winter. The average rainfall is about 500mm. The hot and dry climate support scrubby, throny and xerophytic vegetation. In all 422 plants and 100 species of fauna were recorded from the forest. This is a unique forest, because no were in India forest exists near a city, under virtual threat a extinction.

The total number of species recorded from the reserve forest were 422, belonging to 96 Angiospermic family. The trees were 16.35 per cent and herbs were 57.11 per cent. Leguminoceas and Poaceae were the largest families. The biological spectrum showed maximum Therophytes (51.54 per cent) and Phenerophytes were 25.26 per cent. The quantitative analysis recorded 9 plant associations. Thus the area had rich reserve forest.

The forest is over exploited by indiscriminate felling of trees. The cattle are grazing and destroying saplings. Human beings converting the forest area into a picnic spot and playground. This human gathering produces noise and other types of pollution. As such the natural vegetation has been removed by exotic species or for plantation of mango orchads. The development of housing colonies resulted in stopping seepage of lake water into the forest. The road constructions inside the forest area has polluted the forest by waste material, plastics, air and noise pollution. It was observed that the human settlement near the forest destroy the vegetation and disturb the fauna. The priorities areas to preserve the reserved forest and to prevent its degradation are discussed in the paper.

Keywords: Victoria park, Reserve forest, Human activity, Grassing construction, Priorities areas.

Introduction

Men remain as an integral and inseparable part of nature, since he alone has the capacity to change his own environment and also to alter the environment of other organisms. He plays a pivotal role on ecosystem as his action predominantly affect nature. It is necessary for his own survival and survival of plants and animals that man should preserve the present ecosystem and if possible he should try to improve it. Industrial revolution, though beneficial to the mankind, has left harmful impact on environment some of which are irreversible.

Nature has bestowed upon mankind the most precious gift the forest which is important form many points of view. Forest serves the mankind all basic necessities like food, fuel, gums fibers, resins etc. It helps in bringing rain and also helps in recharging our ground water reservoirs. Forest are the only Sources of oxygen and absorbs Carbon dioxide. It prevent soil erosion reduce the force of torrential rain before it reaches the soil and also give protection against cyclonic wind. Trees in the forest reduce aerial pollution and also helps in maintaining the balance of ecosystem. The forest Department, with the help of peoples participation should take effective steps to extend and conserve forest wealth. "Vanamahotsav" activities improve tree plantation and should be carried out regularly and properly. Conservation of forest becomes an absolute necessity as it is a habitat for a variety of living organisms including wild life.

The Victoria park is reserved forest situated at a distance of 3 km south of Bhavnagar city in Gujarat state. Bhavnagar lies in saurashtra, Gujarat at 72°11' E longitude and 21°45' N latitude and 11 M above sea level. Total area of the forest is 202.74 hectares. The park is triangular in shape. Most of the area in the forest is plain but the western side is hilly and rugged in nature. Roads are constructed on the three sides of the park.

This park was designed under the guidance of Councilor Mr Proctor Sins of the State. The Victoria Park was set up in the year 1888 by Maharaja Takhtasinhji of Bhavnagar. The reserved forest used to house hundreds of species of flora and fauna.

The soil in this forest is forest is of three types. Half decomposed soil just beneath the upper surface called "morrum" forms the first type. The second type of soil is the coarse soil mixed with clay and the third type is the yellowish brown soil. The pH value of soil ranges from 8.45 to 8.95 and hence alkaline cultivation is being carried out in medium black soils.

The ecoclimate of forests are defined in terms of mega-thermal types, depending on climatic conditions. Victoria Park falls in the 'Fourth mega-thermal type" with water deficiency throughout the year. Bhavnagar is a coastal city that experiences semi-arid type of climate with marked seasons. It has very hot summers and cold winters. The region gets very low rainfall.

According to the meteorological and solar radiation data of Bhavnagar the average maximum temperature is 39.8°C and the average minimum temperature is 13.3°C. May is the hottest month of the city and January is the coldest month.The annual average of maximum temperature of the city is 33.6°C and annual average of minimum temperature is 21.4°C. The wind velocity is maximum during the month of June that is 12.1 Km/hr wind velocity is minimum during the month of October, November and December.

Thorny type of vegetation is found in Victoria Park. Thorny, spiny, scrubby and xerophytic types of trees are found in plenty in the forest due to high temperature, low humidity and low rainfall conditions. Initially *Acacia senegal* was the dominant plant species of the park but now due to mismanagement the exotic weed *Prosopis juliflora* DC., appears to have infilterated into the park discouraging the growth of indigenous species. Exotic species like *Prosopis juliflora* DC is found in plenty in the region.

The Victoria Park consists of 422 plants. Some of the common and prominent floral species of the forest area are as follows:

Maytenus emerginata (Wild.) D. Hou. *Acacia nitotica* (L.) Del, *Acacia Senegal* Wild. *Capparis decidua* (Forsk.) Fedgew, *Cassia auriculata* L. and *Capparis sepiaria* L.

Materials and Methods

Several field trips were made at regular intervals to various parts of Victoria Park reserved forest extending over a period of more than ten years (1990 to 2000), for the intensive and extensive collection of plants. For collection and preservation the procedures given by Jain and Rao (1977) and Balgooy (1987) were generally followed. The trips were arranged in such a way so, to as cover all the localities and collected the plants in flowering or fruiting stages. All the specimens collected were serially numbered. The field notes were taken regularly, included habitat, color of the flowers, association and other pertinent features. Efforts were made to identify the plants from the fresh material; those which could not be satisfactorily identified in the field were brought to the laboratory and identified by checking it with monographs, herbarium specimens and other available literature. Collected plants were properly processed, numbered and prepared for herbarium. The herbarium specimens were labeled and deposited in Herbarium Section of the Department of Life Sciences, Bhavnagar University, Bhavnagar.

Floristic Study

Mac (1986) studied flora of Surat district and reported 896 species with five new records in the area. The floristic, phytosociology and ethno botanical study of Vapi and Umargam was studied by Contractor (1987), and reported 964 angiospermic plants. Vashi (1985) worked in the Umarpada forest of Surat district and recorded 751 plants species from the area. Vora (1980) studied the flora of Dharampur, Kaprada and Nana-Ponda ranges. Joshi (1983) made floristic and phytochemical survey of South Gujarat forest. Bedi and Sabnis (1983) studied the ethno botanical aspects of Dadra-Nagar Haveli and Daman region. Rao and Sabnis (1983) studied the floristic and phytochemistry of Kutch and worked on the phytochemical screening of over 600 plants.

The urban ecosystem of Baroda and Surat had been studied by Patil and Sabnis (1982), in great detail to assess the vegetational changes or damages due to industrialization and urbanization. They have pointed out that floristic change due to increasing human activities. Reddy (1987) collected 879 plant species form the Dharampur forest; and also reported 12 species for the first time from the area. Vora and George (1987) studied distribution of various life forms in the ground flora under different canopies of Panchmahal forests. They found that herbs or annual dominated in the distributed areas; while shrubs were seen in the protected areas.

Vora *et al.* (1981) reported 341 flowering plants of Ghoghamahal of Bhavnagar district. The floristic and ecological study of Bhavnagar and its surroundings has been studied by Oza (1991), and reported 528 flowering plant the area. Desai (1992) has recorded five new taxa from Bansda forest, for the first times, which are new additions to the flora of Gujarat State. Pandit and Kotiwar (2002) enumerated 431 species from the Gir forest ecosystem Gujarat, out of which 294 belongs to genera and 94 families of flowering plants. Further they reported 371 dicots and 60 monocots.

Results and Discussions

Total number of species recorded from this forest were 422, belonging to 96 angiosperm families. The ratio of species belonging to monocots and dicots is 1: 4.86; of genera 1: 4.22 and of families 1: 4.65. The ratio of the total number of genera to species is 1: 1.37, which is rather low in comperation to a corresponding ratio for whole of India (1: 7), but it is more or less in conformity with this ratio for the W Rajasthan as reported by Bhandari (1978) and that of Delhi state as reported by Maheshwari (1963).

It is also interesting to note that per cent occurrence of the herbs were 57.11, of shrubs 15.88, of trees 16.35 and of climbers 10.66 per cent.

It is evident that leguminoceae (*sensu lato*) and Poaceae are the largest families amongst the Dicot and Monocot respectively. Families Euphorbiaceae and Asteraceae take up next positions in the area. In order to get an insight into the relations of the flora of Victoria Park with the neighbouring places, a comparative list of the ten dominant families in Victoria Park, Ghogha mahal, Saurashtra region and Gujarat state as whole is given in order of their frequency (Table 3.1).

Table 3.1: Comparison of Ten Dominant Families of
Gir Forest with Different Regions of Gujarat State

Sl.No.	Victoria Park	Ghogha Mahal	Saurashtra Region	Gujarat State
1.	Fabaceae	Fabaceae	Fabaceae	Fabaceae
2.	Poaceae	Poaceae	Poaceae	Poaceae
3.	Euphorbiaceae	Euphorbiaceae	Asteraceae	Cyperaceae
4.	Asteraceae	Asteraceae	Malvaceae	Asteraceae
5.	Malvaceae	Malvaceae	Acanthaceae	Acanthaceae
6.	Acanthaceae	Convolvulaceae	Convolvulaceae	Euphorbiaceae
7.	Convolvulaceae	Acanthaceae	Euphorbiaceae	Malvaceae
8.	Amaranthaceae	Amaranthaceae	Cyperaceae	Convolvulaceae
9.	Cyperaceae	Cucurbitaceae	Rubiaceae	Scrophulariaceae
10.	Tiliaceae	Tiliaceae	Cucurbitaceae	Cucurbitaceae

The biological spectrum reveals a Throphytic plant climate. As already stated, ecoclimate of the area is semi arid, fourth megathermal with no water surplus through the year. Therophytes are best adapted to the general climate of the area and so far most abundant (51.54 per cent). They complete their life cycle within the favorable short period of three months of monsoon and pass the rest unfavorable period of the year in the form of seed. Phenerophytes are present in lesser (25.26 per cent) than on the normal spectrum (46 per cent) and many of them are thorny or spiny, shrubby and xerophytic.

Forest vegetation was analysed for phytosociological characters by quantitative methods and in total 9 associations have been arrived at by an objective method of statistical computation of coefficient of variation of density and cover and finally computing relative growth index (Pandeya *et al.*, 1967).

The associations have been grouped into the following 3 types.

1. Associations with *Acacia Senegal* as the dominant species.
2. Associations with *Prosopis juliflora* as the dominant species.
3. Association mixed in nature but without dominance *of Acacia Senegal* or *Prosopis juliflora.*

There are many factors responsible for the ecological degradation of the forest. The following are the causes discussed in detail.

Victoria Park is the natural forest with plenty of animal, birds and plants species. Initially the forest was dense with abundant trees and underground vegetation. The forest is getting slowly degraded and no significant steps are taken to offset the damage.

Victoria Park is surrounded by walls on all its sides. At some places, the edges are broken which allows cattle to enter the park. Cattle, goat and sheep enter the park for grazing and destroyed the plant seedlings.

There are many houses close to the park. Human settlement near the park destroy the vegetation and disturb the fauna. People cut trees for domestic purposes.

People use this place as a garden for recreation. Unfortunately people who come here pollute the park by dropping tibbits and waste materials. Large human gathering usually produce noise pollution, which is not a conducive factor for the fauna. In this area, natural vegetation is now being removed and the area is now converted into a park, thereby deteriorating the forest area.

Actually thorny, scrubby and spiny plants which exist in large abundance is ideally suited for the preservation of the Victoria Park. Indiscriminate mango plantation in the forest, degrading the thorny trees has proved a serious threat to the very existence of the park.

Another major cause for ecological degradation is the policy decision of forest department, who favour a mixed type over monoculture vegetation. It should be mentioned here that the forest at present includes xerophytic plants and mango plants between them. Only the indigenous plants of the park should be planted and harmful weeds and cuscuta parasites should be eradicated.

At a few places the natural vegetation of the forest has been removed and exotic species like *Prosopis juliflora* exists.

Human activities near the Krishnakunj lake disturb the migratory birds. Housing development is encouraged near the lake. Well-grounded apprehensions are expressed that the housing activity would result in stopping of seepage of lake water thus ending the much necessary habitat and breeding ground for wetland birds. People in these areas depend on the water form this lake thereby reducing the availability of water for animal and birds. Silting has made the lake shallow, hence the silt should be dredged out so that the lake can hold more water.

Transport passing through the roads and people walking on these roads pollute the lake by dropping waste materials and other effluents. People burn the twigs of the forest for cooking purposes. The smoke generated affects the leaves of the plants. The dust particles produced gets deposited on the surface of the leaves reducing their photosynthetic ability.

Rainfall is unevenly distributed affecting the vegetative growth. Heavy down pour of rain water uproots the trees and degrades the forest.

The Victoria park a unique gift to the people of Bhavnagar city is under a virtual threat of extinction. Once upon a time the forest was endowed with wild boars, cheetals, hog deer and chinkaras giving a rare glimpse to nature lovers. Teak, bamboo, rudraksha and sandal wood trees were introduced in few numbers. In all 422 plants and 400 birds were observed in the forest, due to the various causes mentioned above the natural beauty of the forest is destroyed, the number of flora and fauna species is getting reduced.

Victoria Park is one of the most beautiful park found near the Bhavnagar city. The priorities to preserve the park and to prevent its ecological degradation the following steps can be adopted.

The forest department of Bhavnagar city must take the first step to prevent the natural degradation of the park. Human settlement in and around the park may be avoided.

Preventive measures may be adopted for, (*i*) collection of dry wood from the forest, (*ii*) cutting down trees.

Steps should be taken to slowly replace the existing picnic spot by planting trees, removing the playing material so that the ecological balance is restored.

The water of the Krishnakunj lake in the Victoria Park is situated for the natural flora and fauna of the park. Using of this water for human necessities should be objected. Reservoirs are constructed across the lake for diverting the water of the lake for drinking and sanitation purpose. Instead of using the water for human settlement, it should be supplied to plants in the forest. As there are many reptiles, fishes and amphibians in the forest sufficient amount of the water should be present in the lake.

Arrangements should be made to conserve the rainwaters. The original thorny vegetation of the forest should be practiced thoroughly instead of adopting new mango plantations. Nurseries should not be encouraged in side the park.

Roads should not be constructed in the park. People should improve the natural scenic beauty of the forest instead of destroying it by increasing the number of plants and animal species. The entire area should be declared as "PROTECTED AREA" and strict watch should be maintained on the trespassers.

The entry could be by way of issue of tickets of nominal value so that the funds so collected can be utilized for development of forest.

Short films through slides and projectors should be shown to the people either free of charge (or) by charging a nominal fees highlighting the importance of preserving the forests.

Thus, if such steps are taken for this beautiful forest it can be protected by which we would be doing justice to the natured, greatest gift to mankind.

In general, by protecting the existing forests and also by taking effective steps to plant new trees ecological balance of this beautiful planet can be maintained and man can lead a trouble free life.

Let us all think and act in this directions.

Conclusions

Forests are managed for a variety of purposes, *viz.*, timber for industrial use, forest products for rural communities, watershed of multipurpose use, and land use alterations based on land capability for carbon sequestration and/or conservation of biodiversity. Each objective requires a different approach to management, but some generalizations can be made.

 ☆ Management can be succeed only if economic benefits are assured to local communities.
 ☆ Women's participation is necessary.

☆ Total understanding of the complexities of the system.

☆ Management should lead to improvement of soil fertility and water quality.

☆ Resources recycling within the system.

☆ Self-regenerating capacities are to be enhanced.

☆ Strong community participation and ownership of participants should be well informed.

☆ User group/Community

☆ In semiarid region ground water resources need to be monitored.

☆ Establishment of energy farm for local communities.

In conclusion, it would be better to end with the statement given by Odum (1971): "We can now make a strong case for the proposition that adequate pollution free living space, not food, should be the key to determining the optimum density for man. In other words, the size and quality of the "ecos" or "environmental houses", should be the limiting consideration, not the number of calories we can relentlessly squeeze form the earth. A reasonable goal is to make certain that least a third of all land remains in protective open space use. This means that such a portion of our total environment should be in national, state or municipal parks, ownership, it should be protected by scenic easements, zoning or other definitive legal means".

References

Bedi, S.J. and Sabnis, S.D., 1983. Ethnobotany of Dadra-Nagar Haveli and Daman. *Ind. J. For.*, 6(1): 65–69.

Bhandari, M.M., 1978. *Flora of the Indian Desert*. Scientific Pub., Jodhpur.

Contractor, C.J., 1987. Floristic, phytosociology and ethnobotanical study of Vapi and Umargaon area in South Gujarat. *Ph.D. Thesis*, S.G. University, Surat.

Desai, M.J., 1992. New records for Gujarat flora from Bansda forest. *J. Econ. Tax. Bot.*, 16(3): 551–552.

Joshi, M.C., 1983. A floristic and phytochemical survey of some important South Gujarat forests with special reference to plants of medicinal and ethnobotanical interest. *Ph.D. Thesis*, M.S. University, Baroda.

Pandit, B.R. and Kotiwar, O.S., 2002. Floristic composition of tropical dry deciduous (Gir) forest ecosystem. In: *Emerging Areas in Plant Sciences*. Bhavnagar University, Bhavnagar, pp. 111–132.

Mac, R.N., 1986. A contribution to the flora of Surat district. *Ph.D. Thesis*, S.G. University, Surat.

Maheshwari, J.K., 1963. *The Flora of Delhi*. CSIR, New Delhi.

Odum, E.P., 1971. *Ecology*. Modern Biological Series, Holt, Renehart and Winston, New York.

Oza, R.A., 1991. Taxonomical and ecological study of the flora of and around Bhavnagar. *Ph.D. Thesis*, Bhavnagar University, Bhavnagar.

Pandeya, S.C., Pandya, S.M., Murthy, M.S. and Kuruvilla, K., 1967. Forest ecosystem: Classification of forest vegetation with reference to forest in river Narmada catchment area. *J. Indian Bot. Soc.*, 46(4): 412–427.

Patil, S.N. and Sabnis, S.D., 1982. New plants from urban environment of Baroda, Gujarat. *J. Bom. Nat. Hist. Soc.*, 79(1): 117–119.

Reddy, A.S., 1987. Flora of Dharampur forest. *Ph.D. Thesis,* S.P. University, Vallabh Vidhyanagar.

Vashi, B.G., 1985. Florestic, phytosociology and ethnobotanical study of Umarpada forest in South Gujarat. *Ph.D. Thesis,* S.G. University, Surat.

Vora, A.B. and George, V.C., 1987. The distribution of various life forms in the ground flora under different canopies of Panchmehal forests, Gujarat. *Ind J. For.*, 10(3): 223–225.

Vora, H.M., 1980. A contribution to the flora of Dharampur, Kaprada and nanapondha ranges. *Ph.D. Thesis,* S.G. University, Surat.

Vora, U.A., Patel, B.P. and Patel, B.K., 1981. The vegetation of Ghogha mahal and its biological spectrum. *Geobios,* 8: 211–214.

Chapter 4

Plant Diversity in India: An Overview

R.N. Trivedi

Department of Forest Ecology, Biodiversity and Environmental Sciences, Mizoram Central University, Aizawl – 396 007, Mizoram

Introduction

India, a mega-biodiversity centre, while following the path of development, has been sensitive to the needs of conservation. More than 30 per cent of the species occurring in the Indian-subcontinent are endemic. The Western Ghats, North-Eastern region and Western Himalayas are considered as three mega-centres of endemism in Indian flora. There are 26 such areas, which have been declared as Microcentres of endemism. The great geographical expanse of the country has resulted into enormous ecological diversity. India occupies about 2.4 per cent of world's total landmass but harbors 45,000 plant species (Karthikeyan 2000) out of which about 0.4 million species hitherto known in the world and has about 11 per cent of the world flora. The diversity ranges from sea level to the highest mountain ranges in the world and the climate varies from hot and arid bearing dry deciduous forests in the west to warm and humid in the North Eastern region and western ghats with tropical evergreen forest and cold arid conditions in the Trans-Himalayas to fresh water aquatic to marine ecosystems of Sunderbans. This wide variation in geographical, climatic, topographical etc. conditions have resulted in several types of forests, grassland, wetlands, coastal, marine and desert ecosystem each with rich biodiversity characteristics of its own (Sharma and Singh, 2001). Champion and Seth (1968) have recognized 16 major forest types comprising 221 subtypes. In view of their management and use, Ministry of Environment and Forests, Government of India recently has launched a project *viz.* National Biodiversity Strategy and Action Plan (NBSAP) which envisages the assessment and stock taking of biodiversity related information at various levels.

Biological diversity constitutes resource upon which humanity depends. It is not uniformly distributed on the earth. Increasing human interventions on the ecosystems have accelerated the process of biodiversity loss. Though process of speciation and extinction is part of evolutionary process, it is reported that the speciation is more concentrated along tropics where the ecological set up and continental readjustments have permitted higher rate of speciation. The biodiversity dynamics analysis reveals that the species extinction process has overtaken speciation process. India's strategies for conservation and sustainable utilization of biodiversity in the past have been comprised by providing special status and protection to biodiversity rich areas by declaring them as national parks, wildlife sanctuaries, end biosphere reserves, ecologically fragile and sensitive areas. It has helped in off loading pressure from reserve forests by alternative measures of fuel-wood and fodder need satisfaction by afforestation of degraded areas and wastelands, creation of *ex situ* conservation facilities such as gene banks and ecodevelopment. The challenges before India are not only to sustain the efforts of the past but also further add to these efforts by involving people in the mission.

An environment, which is rich in biological diversity, offers the broadest array of options for sustaining human welfare and for adapting to change. Loss of biodiversity has serious economic and social costs for any country. The experiences of the past few decades have shown that as industrialization and economic development take place, the patterns of consumption, production and needs change, strain, alert and even destroy ecosystems. If one traces the linkages of biodiversity, it is observed that human population and its interventions have been the major factors, responsible for the extinction of species. Conservation and sustainable use of biodiversity, hence, is fundamental to the sustainable development. The mapping of mega diversity zones indicates major concentration in South America, Central Africa and India/South China.

India is also known as one of the world's 12 Vavilovian centres of origin and diversification of cultivated plants known as the Hindustan Centre of Origin of Crop Plants (Vavilov, 1951). Biodiversity has direct consumptive value in agriculture, medicine and industry. Vavilov estimated that about 80000 edible plants have been used at one time or the other in human history of which about 150 have even been cultivated on a large scale. Today merely 10 to 20 plant species provide 80–90 per cent food requirements of the world. In India, rural communities, particularly the tribals obtain considerable part of their daily food from wild plants. At one time nearly all medicines were derived from biological resources. Around 20000 plant species are believed to be used for medicine in the developing world. In India, the knowledge about medicinal value of plants has evolved in the form of traditional systems of medicinal sciences like, Unani, Ayurveda and Siddha. More than 8000 species are used in some 10000 drug formulations. The source of ninety percent of Industry's requirements is from forests. It is estimated that about 0.5 million ton (dry weight) of plant material is collected each year from the forests and often the method of collection is destructive. The global plant based drug trade is projected around US $ 62 billion with a 7 per cent annual growth rate but Inida has only 2.5 per cent share in it. The outstripping supply has put tremendous pressure on our wild phytoresources which is resulting in the erosion in biodiversity. About 30 per cent of the area in the country is unexplored or under-explored which houses a number of species yet to be discovered (Sharma and Singh 2001).

Levels of Biodiversity

Biodiversity is the totality of genes, species, ecosystems and habitats in a region. Hence, the best way to understand biodiversity is to look at it in a hierarchical manner in which biological organisms are organized. The smallest unit starts from the diversity contained in the genetic material of individual organisms and then goes on to encompass the biological communities in which species are organized and on to the ecosystems in which they exist. Biodiversity is recognized at three levels *viz.*, genetic, species and ecosystem levels. At all three levels of biodiversity the components are dynamic in space and time.

The genetic composition of species changes over time in response to natural and human induced selection pressures, occurrence and relative abundance of species in ecological communities as a result of ecological and physical factors. The ecosystems at the same time strongly respond to external dynamics and internal stresses. Ecological equilibrium is important for species to occur, grow and evolve through the natural processes.

Protected Area Concept in Biodiversity Conservation

A protected area is defined by the Convention on Biological Diversity as a geographically defined area, which is designated or regulated and managed to achieve specific conservation objectives. The land in the protected areas has certain legal provisions, which facilitates the management of protected areas. At present about 4.66 per cent of total geographical area of the country is under different kinds of protection. This includes 86 National Parks and 480 Wildlife Sanctuaries (Rodgers *et al.*, 2000). There are 12 biosphere reserves in the country other than protected and reserved forests.

Explicit biodiversity conservation objectives need to be established for each protected area and in most cases, they need to be better integrated into the fabric of social, environmental and economic welfare. If protected areas are to become more effective in maintaining biodiversity, serious obstacles must be overcome which include:

☆ Inadequate biogeographic distribution;

☆ Definition of arbitrary boundaries;

☆ Conflict with local people;

☆ Ineffective management and funding; and

☆ Limited appreciation of potential roles in sustainable development.

It is observed that the process of selecting and classifying protected areas in the country was arbitrary, unsystematic and inconsistent. Protected areas were also not

Advantages: Precision as a measure of Character Diversity	A scale of surrogacy of Character Diversity	Advantage: Inexpensive survey and units more Inclusive of viability enhancing process
LOW Environmental surrogates	Ecosystem richness? Climate class richness	HIGH
Environmental/ Assemblage surrogates	Terrain class richness Substrate class richness	
	Landscape class richness	
Assemblage surrogates	Habitat class richness Community class richness	
Taxonomic surrogates	Vegetation class richness	
	Higher taxon richness Species/sub species richness	
Molecular surrogates HIGH	Taxonomic/Phylogenetic subtree length Expressed gene richness	LOW

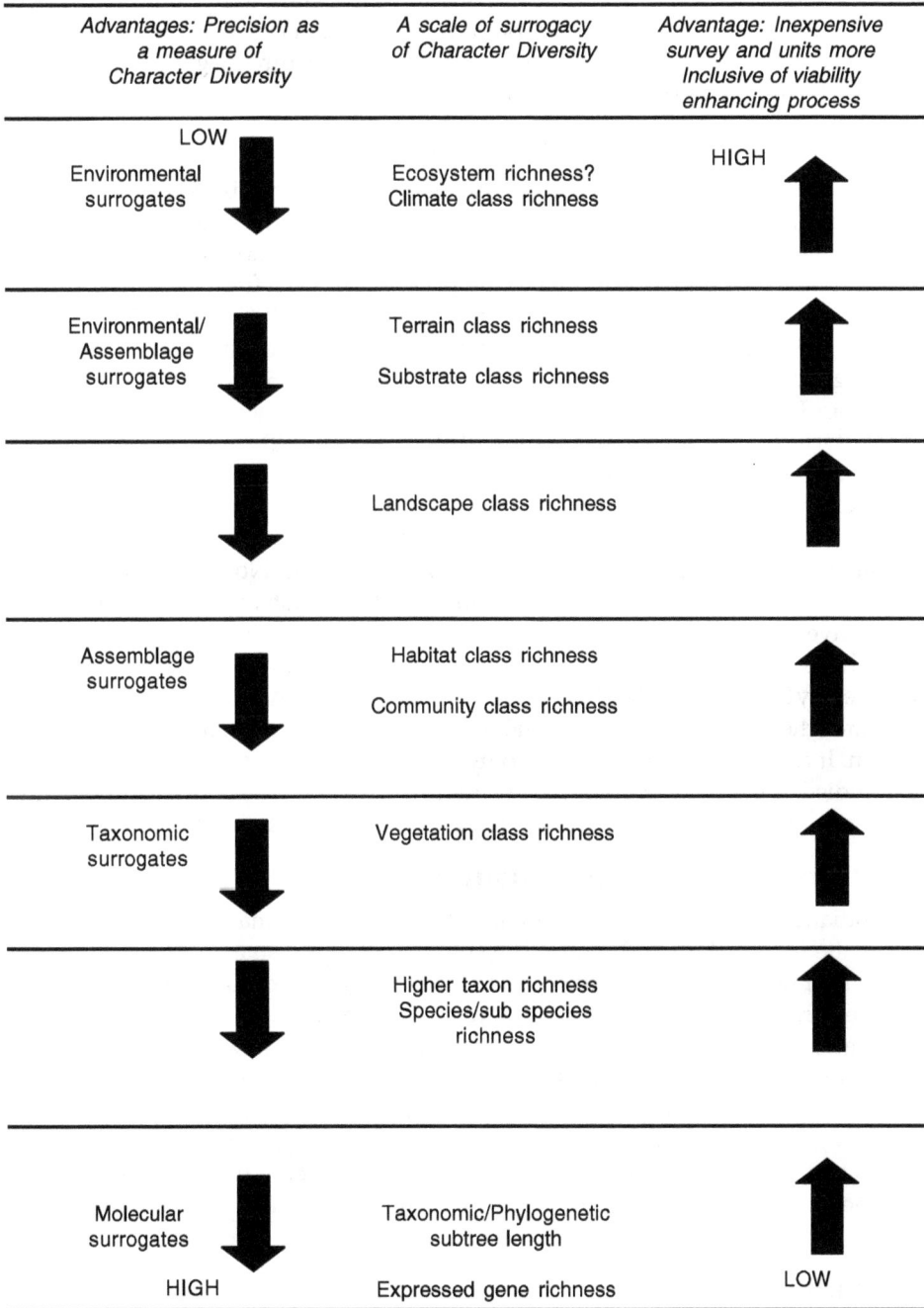

Figure 4.1: Surrogates in Measuring Biodiversity Value
(*Source*: Paul Williams 2000)

placed in any rational system of regional land use planning. As a result, even densely settled areas were designated as national parks and many important biodiversity areas were not included in the network. As many as 60 per cent of national parks and 92 per cent of the sanctuaries have not completed the required legal procedures for their establishment as protected areas.

Global Initiatives for Biodiversity Assessment

The initiative for biodiversity assessment was taken long back in 1991 with the UNEP Biodiversity Country Studies Project (consisting of bilateral and Global Environmental Facility funded studies in developing countries) implemented in cooperation from donor countries and UNDP. The approach from gene to ecosystem was initiated as a Research Agenda for Biodiversity IUBS/SCOPE UNESCO. Paris (Solbrig, 1991). The agreed text of the Convention Biological Diversity was adopted by 101 governments in Nairobi in May 1992. Signed by 159 governments and the European Union at the Nations Conference on Environment and Development (UNCED) held at Rio de Janeiro in June 1992. At present 120 governments are party to this convention. Apart from this Global Biodiversity Strategy (1992), Global biodiversity: Status of the Earth's Living Resources (1992), Caring for the Earth: A Strategy for Sustainable Living (1991), Global Marine Biological Diversity: A Strategy for Building Conservation in to Decision Making (1993), Norway/UNEP Expert Conference on Biodiversity (1993) and From Genes to Ecosystems: A Research Agenda for Biodiversity (1991) are the milestones on the international biodiversity initiatives. More than scores of nations and regions are engaged in developing their own National Biodiversity Strategies or Action Plans. Global Biodiversity Assessment (UNEP, 1995) estimates the total number of animal and plant species to be between 13 and 14 million. It further records that so far only 1.75 million species have been described and studied. Ecosystem diversity has not been even reasonably explored as yet. Hence, there seems to be wide gap of knowledge at global, regional and local level.

Biodiversity Assessment: Initiatives in India

Indian sub-continent is also known for its diverse bioclimatic region supporting one of the richest flora and fauna. The sub-continent is also a confluence point of three major terrestrial biogeographical realms (*viz.*, the Indo-Malayan, the Eurasian and the Afro tropical) and the Antarctic realm for the marine biodiversity. In a most recent attempt to map Biogeographical regions, Rogers and Pawar (1988) attempted to define the biogeographical regions of India. The subcontinent has ten biogeographical zones *viz.*, Trans-Himalayan, Himalayan, Indian desert, Semiarid, Western ghats, Deccan peninsula, Gangetic plains, North East India, Islands and Coasts and not yet defined zones for Aquatic (freshwater and marine) ecosystems have been mapped. The Wildlife Institute of India has converted these regions on Survey of India digital database.

Measuring Biodiversity

In recent years we have been confronted with a situation wherein we need to know the methods to measure biodiversity, so that we can determine "where" *in-situ* conservation action is required: rather than "how" to implement the site specific biodiversity conservation plans. The latter is ground-based and enough scientific options are available. Botanical Survey of India for flora and Zoological Survey of India for fauna are premier agencies in the country for inventoring biodiversity. They follow traditional methods of data collection, storage and listing. Remote sensing can play an important role in prioritizing areas for explorations.

Biodiversity can be seen as irrefutable complexity of all life. The objective measure of biodiversity is a difficult proposition. However, measures relative to some particular purpose can be applied. For conservationists, the measure of biodiversity should quantify a value shared broadly, among the people for whom they are acting. One such value is to ensure continued possibility for adaptation and use by people of changing world. Arguments for measuring biodiversity value as character richness at least provide a reasonable starting point, as one possible answer to the question of what is valued in diversity. The remote predictors or surrogates often play very significant role to measure richness. Details on the physical environment, factors determining the biodiversity loss in a spatial context may be of practical information value and could reduce sampling intensity. This information base could also guide detailed sampling on the ground. These large scale surrogates include entire functional system and are more likely to promote population viability in the ecosystem.

We need to be very clear with respect to the biodiversity measures. In conservation this is likely to differ with earlier measures of ecological diversity formulated with the narrower aim of representing differences in abundance among species, exploring distribution of resources within community. If the value of biodiversity to a conservationist, is associated with its use to people then this ought to be separated carefully from issues of rarity, viability and threat. If the biodiversity value is associated with richness in a currency of characters of organism then the higher level of biological organization (or environmental factors affecting its distribution) will have to be used as surrogate measures. Choosing a surrogate level from this scale is a compromise between the precision of the measure on the one hand, the availability and cost of data compilation on the other. Higher level surrogates should have the additional advantage of implicitly integrating more of the functional processes that favours viability. The taxonomic inventories in the past have only been able to reach partial level of understanding the richness. Hence, this should be a 'top down' and 'bottom up' approach together.

References

Anonymous, 1983. *Flora and Vegetation of India: An Outline*. Botanical Survey of India, Howrah, pp. 2331–2346.

Belal, A. and Springuel, I., 1996. Economic value of plant diversity in arid environments. *Nature and Resources*, 31(1): 33–39.

Champion, H.G. and Seth, S.K., 1968. *Revised Forest Types of India*. Govt. of India Publications, New Delhi.

Chatterjee, D., 1939. Studies on the endemic flora of India and Burma. *J. Asiat. Soc. Bengal*, 115(3): 19–67.

Forman, T.T.R. and Gordon, M., 1986. *Landscape Ecology*. Wiley and Sons, New York.

Hajra, P.K., Verma, D.M. and Giri, G.S., 1996. *Materials for Flora of Arunachal Pradesh, Vol–1*. Botanical Survey of India, Calcutta.

Khan, M.L., Menon, S. and Bawa, K.S., 1997. Effectiveness of the protected area network in biodiversity conservation: A case study of Meghalaya. *Biodiversity and Conservation*, 6: 853–868.

Khoshoo, T.N., 1993. Biodiversity in India: Reality and myths. In: *Science and Technology for Achieving Food, Economic and Health Security*, (Ed.) U.R. Rao.

Kothari, A., Pandey, P., Singh, S and Variava, D., 1989. *Management of National Parks and Sanctuaries in India: A Status Report*. Environmental Studies Division, Indian Institute of Public Administration, New Delhi.

Nagendra, H. and Gadgil, M., 1999. Biodiversity assessment at multiple scales: Linking remotely sensed data with field information. In: *Proceedings National Academy of Sciences, USA*, 96: 9154–9158.

Nayar, M.P. and Shastry, A.R.K. (eds.) 1987. *Red Data Book of Indian Plants, Vol.1* Botanical Survey of India, Calcutta.

Nayar, M.P. and Shastry, A.R.K. (eds.) 1988. *Red Data Book of Indian Plants, Vol. 2*. Botanical Survey of India, Calcutta.

Nayar, M.P. and Shastry, A.R.K. (eds.) 1990. *Red Data Book of Indian Plants, Vol. 3*. Botanical Survey of Indian, Calcutta.

Ramesh, B.R., Menon, S. and Bawa, K.S., 1997. A vegetation based approach to biodiversity gap analysis in the Agastyamalai region, Western Ghats, India. *Royal Swedish Academy of Sciences*, p. 529–536.

Rao, R.R., 1994. *Biodiversity in India: Floristic Aspects*. Bishen Singh Mahendra Pal Singh, Dehradun.

Rodgers, W.A., Panwar, H.S. and Mathur, V.B., 2000. *Wildlife Protected Areas Network in India: A Review*. Wildlife Institute of India, p. 1–44.

Romme, W.H., 1982. Fire and landscape diversity in sub-alpine forests of Yellowstone National Park. *Ecol. Monogr.*, 52: 199–221.

Roy, P.S. and Tomer, S., 2000. Biodiversity characterization at landscape level using geospatial-modeling technique conservation, 95(1): 95–109.

Roy, P.S., Singh, S. and Porwal, M.C., 1993. Characterization of ecological parameters in tropical forest community: A remote sensing approach. *Journal Indian Society of Remote Sensing*, 21(3): 127–149.

Roy, P.S., Tomar, S. and Jegannathan, C., 1997. Biodiversity characterization at landscape level using satellite remote sensing. *NNRMS Bulletin*, B–21: 12–18.

Sharma, J.R. and Singh, D.K., 2001. Status of plant diversity in India: An overview. In: *Proceedings Biodiversity and Environment*, Indian Institute of Remote Sensing, Dehradun.

Solbrig, O.T. (ed.), 1991. *From Genes to Ecosystems, A Research Agenda for Biodiversity,* IUBS/SCOPE/UNESCO, Paris.

Vavilov, N.I., 1951. *The Origin Variation Immunity and Breeding of Cultivated Plants.* Selected Writings. Ronald Press, New York.

Whittaker, R.H., 1972. Evolution and measurement of species diversity. *Taxon,* 21: 213–251.

Whittaker, R.H., 1977. Evolution of species diversity in land communities. *Evolutionary Biology,* 10: 1–67.

Chapter 5

Ethnomedicinal Importance of Some Common Pteridophytic Plants Found in Aravalli Ranges of Rajasthan

Pradeep Parihar[1]* and A. Bohra[2]

[1]Department of Biotechnology, Lovely School of Technological Sciences, Lovely Professional University, Phagwara, Punjab – 144 402
[2]Microbiology Laboratory, Botany Department, J.N.V. University, Jodhpur – 342 005, Rajasthan

Pteridophytes (fern and fern allies) by virtue of their possessing great variety and fascinating foliage have drawn the attention and admiration of horticulturists and plants lovers for centuries. They are represented by about 305 genera, comprising more than 10,000 species all over the world. About 191 genera and more than 1000 species are reported from India (Dixit, 1984; Bir, 1992). Medicinal value of pteridophytes is known to man for more than 2000 years. Theophrastus (Ca. 327-287 B.C.) and Dioscorides Ca. 100 A.D.) have referred to medicinal attributes of certain ferns. Sushruta and Charaka (Ca. 100 A.D.) mentioned medicinal uses of *Marsilea minuta* Linn. and *Adiantum capillus-veneris* Linn. in their samhitas. Kirtikar and Basu (1935) described 27 species of ferns of which only 19 are used in India. Chopra *et al.* (1933 and 1956) and Nadkarni (1954) have referred to 44 and 11 species of

* Corresponding Author E-mail: pradeepparihar2002@yahoo.com.

pteridophytes of medicinal importance respectively. However, Jain and De Philipps (1991) concluded that only 31 species of pteridophytes are covered in books dealing with medicinal plants in India.

Though recent ethnobotanical, phytochemical and pharmacological researches have reported the medicinal and pharmaceutical values of some more species of pteridophytes, still more species of pteridophytes, which are used by the tribal of India are yet to be explored for their pharmaceutical value and to isolate the active principle. Therefore, detailed phytochemical and pharmacological studies will help to evaluate their medicinal attribute and to isolate the active principle. Plants have been one of the important sources of medicines even since the dawn of human civilization. In spite of tremendous development in the field of allopathic during the 20th century, plants still remain one of the major sources of drugs in modern as well as traditional system of medicine throughout the world. Approximately one-third of all Pharmaceuticals are plant-based products, wherein fungi and bacteria are also included. Man from prehistoric times for relieving suffering and curing aliments has used plants. Primitive people, when injured in battle or when they had a fall or cut, used plant materials available at hand for staunching the flow of blood or relieving of pain and, by trial and error, they learnt that certain plants were more effective than others for these purposes. Man has also gained such knowledge from birds and animals, which are using plants for curing their ailments. Even today, we find that the domestic dog and cat, when they suffer from indigestion or other ailments, run to the field, chew some grasses or herbs and get cured. The folk medicines of almost all the countries of the world abound in medicinal plants and tribal people wherever they exist, rely chiefly on herbal medicine, even today.

Today, chemical and pharmaceutical investigations have added a great deal of status to the use of medicinal plants by revealing the presence of the active principles and their actions on human and animal systems. Investigations in the field of Pharmacognosy and pharmacology have been supplied valuable information on medicinal plants with regard to their availability, botanical properties, methods of cultivation, collection, storage, commerce and therapeutic uses. All these have contribute towards their acceptance in modern medicine and their inclusion in the pharmacopoeias of civilized nations.

The indigenous systems of medicine practiced in India are based mainly on the use of plants. *Charaka samhita* (1000 BC–100 AD) records the use of 2000 vegetable remedies. Ancient medicine was not solely based on empiricism and this is evident from the fact that some medicinal plants, which were used in ancient times, still have their place in modern therapy. Thus, for example, *'Ephedra'* a plant used in China 4000 years ago is still mentioned in modern pharmacopoeias as the source of an important drug, ephedrine. The plant Sarpagandha *(Rawvolfia serpentina)* which was well known in India as a remedy for insanity has now shown that one of its constituents, reserpine, is a wonder drug today for curing mental ailments. Quinine, another important antimalarial drug of modern medicine, was obtained from the Bark of cinchona tree. The knowledge about the use of medicinal plants has been accrued through centuries and such plants are still valued even today, although synthetics, antibiotics, etc. have attained greater prominence in modern medicine. It is, however, a fact that these synthetics and antibiotics although they often show miraculous and often instantaneous results, prove harmful in the long run and that is why many synthetics and antibiotics have now gone out of use or have been specified to be prescribed strictly under medical supervision. In the case of most medicinal plants, however, no such cumulative derogatory effect has been recorded and that is why many of the medicines obtained from plants are still widely used today.

The age-old Indian system of medicine, have been neglected mainly because of the rapid expansion of the allopathic system of medical treatment. This is despite the fact that our country has a long history of local health traditions, which are backed by thousands of scriptures left behind by practitioners of these systems of medicine. Over 7000 different species of plants found in different ecosystem are said to be used for medicinal purpose in our country. The traditional Indian system of medicine can be broadly classified into the empirical forms of folk medicines, which are village-based, region-specific, indigenous herb-based, local resource based, and, in many cases, community-specific. The other system called the *Shastriya* stream, which includes the Ayurveda, Siddhaa and Unani systems of medicine, is said to be more complicated and elaborate with theoretical and research findings. It is also said to be documented in thousands of regional manuscripts. About 1,00,000 medical manuscripts are said to be available in oriental libraries and private collection in India and abroad, but only a few hundreds of these books are made available to students and teachers of Indian medicine. Presently, the Indian system of medicine uses over 1,100 medicinal plants and most of them are collected from the wild regularly, of which over five dozen species are said to be in great demand. The tribal belt of India is rich in these plants and local tribes mainly depend, for their livelihood, on their collection and trade.

According to many researchers medicinal plants are expected to increase the immunity power of human body there by preventing further susceptibility to diseases. There is various advantage of herbal therapeutics over the synthetic medicines. A large number of drug plants have been found in India. In India climatic conditions varies from torrid to frigid zones and temperate plains, hills and valleys, irrigated lands, moist and dry climates. India should be termed as the 'Botanical Garden' of the world and the records reveals that about more than 2000 types of medicinal and essential oil bearing plants are present in India. Pharmacopoeia records nearly 1000 medicinal plants and their preparations. Various kind of Indigenous plants have a good market value in exports, like *Senna, Catharanthus, Dioscorea, Digittalis, Rauwolfia, Plantago, Opium* and many others. Keeping in view of these advantages the imperative need of ethnological study of medicinal plants has been attracted the attention of modern workers to study the various aspects of these drugs, their physiological impact on human health, their pharmacological and microbiological aspects.

Now-a-days, the study of the drugs and drug plants has progressed steadily and at present pharmacology is the essential branch of medicine, and Botany and medicine have gone hand in hand and the majority of Botanists of past had a knowledge of medicinal plants. Although antimicrobial properties of the drugs are not mentioned in early literature but therapeutically properties of drugs may be due to presence of chemical substances. Some of which either individually or collectively may be effective as antimicrobial for both gram positive as well as gram-negative bacteria, fungi, actinomycetes, protozoa etc.

India is profusely rich in the history of medicinal plants and its 75 per cent folk population is still using herbal preparations in the form of powder, extracts and decoction because these are easily available in nature and the natives have stronger faith on traditional knowledge. Ministry of health and family welfare center and state government are conducting high-level research programmes to manufacture drugs. These drugs of medicinal value are competing today in markets. Various plants exhibit various types of antimicrobial activities. Orchids also are of great medicinal value. Even parasitic plants were also found to be antimicrobial.

Now various workers have given a considerable attention on antimicrobial properties of various plants. Chopra (1933) screened various indigenous drug plants and their medicinal and economical aspects.

Cain (1935) studies the medicinal and poisonous ferns of India. Sen and Nandi (1951) have isolated an antibiotic from the pteridophytes. Nayer (1958) studied the medicinal ferns of India and enumerate medicinally used pteridophytes of India. Puri and Arora (1961) and (1970) found some medicinal ferns from Western India used in folk remedies to cure various diseases. Lall *et al.* (1964) found some ferns inhibiting *Cercospora species* from India. Dhar *et al.* (1968) screened the Indian ferns for their biological activities against various pathogens. Bhakuni (1969) screened the Indian plants for their biological activity against microorganisms. Egawa *et al.* (1972) studied various antimicrobial substances. Khairsagar and Mehta (1972) surveyed the ferns in Gujarat State for presence of their antimicrobial substances. Kumar and Roy (1972) studied some medicinal ferns from Netharhat hills (Bihar). Singh (1973) isolated some medicinal ferns of Sikkim, Himalayas. Dixit (1974, 1975) studied and published his work on fern–A much-neglected group of medicinal plants in three volumes. Smith and Vee (1975) observed the effect of coconut milk on the germination of spores of *Nephrolepis hirsutula*.

Horsely (1977) studied the allopathic inhibition of black cherry by ferns, grass, golden rot and asters. Kapur and Sarin (1977) isolate some useful medicinal ferns from Jammu and Kashmir. Dixit, Das and Kar (1978) studied the ethnobotany of some less known edible, economic and medicinal ferns of Darjeeling District, West Bengal. Kasmi and Trivedi (1978) studied the antifungal property of some common plants. Banerjee and Sen (1980) have reported that fern extract prove toxic to certain fungi and shows antifungal properties. They also showed that ferns exhibit antibiotics properties, too. Devi, Jamil and Verma (1980) studied the allergic manifestations of fern spores as potential aeroallergens. Pandey and Bhargava (1980) have showed that extract of some pteridophytes also proved to be having antiviral activities. Vicknova *et al.* (1982) studied the antimicrobial activity of some *Dryopteris* sps. from the Primorrk territory. Khandelwal, Gupta and Kaushik (1983) studied the antimicrobial activity of oil of *Ophioglossum*. Bhardwaja and Garg (1984) studied the antifertility effect of an Australian species of the aquatic fern *Marsilea* L. San Fransisco and Cooper (1984) studied antimicrobial activity of phenolic acid in

Pteridium aquilinum. Srivastava and Kediyal (1984) screened the effect of fern extract on conidial germination and germ tube growth of two pathogenic fungi. Stetsenko *et al.* (1984) found antimicrobial properties of introduced ferns. Singh and Roy (1986) studied some medicinal ferns from the Mirzapur forests. Sharma and Vyas (1987) found various fern plants treated to cure various diseases. They studied ethnobotany of ferns and fern allies of Rajasthan.

Mahran *et al.* (1990) studied the chemical composition and antimicrobial activity of volatile oil and extracts of fronds of *Adiantum capillus-veneris* L. Ganesan (1993) studied the effect of leaf extracts of some pteridophytes on *Drachslera oryzae* conidial germination. Paulo, Dauarte and Gomes (1994) have done *in vitro* antibacterial screening of *Cryptolepis sanguinolinta* alkaloids. Gehlot, Bohra and Bohra (1995) have found out the antibacterial activities of leaf extracts of some ferns from Pachmarhi Hills. They also screened some ferns for their antifungal activity. Kaushik and Dhiman (1995) screened out some common medicinal pteridophytes.

Hansraj (1996) found out the medicinal value of *Adiantum.* Paulo *et al.* (1997) studied the chemical and antimicrobial studies on *Cryptolepis obtusa* leaves. Parihar and Bohra (2000) have shown the effect of *Marsilea minuta* on the growth of *Escherichia coli.* Effect of plant part of seven pteridophytic plants was studied against human pathogenic strain of *Escherichia coli by* Parihar and Bohra (2002). Besides antibacterial activity, antifungal activity was also studied by Parihar and Bohra (2002) against *Candida albicans.* Antibacterial efficacy of some pteridophytic plants found in Rajasthan was studied against human pathogenic strain of *Staphylococcus aureus* by Parihar and Bohra (2002). Screening of some ferns for their antibacterial activity against *Salmonella typhi* was also studied by Parihar and Bohra (2002). Anticandidal activity was also studied by Parihar and Bohra (2002).

To find the mechanism of pharmacognosy of various diseases, researchers have found out the phytochemistry of ferns. But very little work has been done on the fern and fern allies of Rajasthan.

Chatterjee *et al.* (1963) studied the chemistry and pharmacology of *Marsilea* and found a sedative and anticonvulsant principle from it and called it as *Marsiline.* It was isolated from *Marsilea minuta* L. and *M. rajasthanensis.* Gupta, Shrivastava, Bhakuni and Sharma (1963) investigated the chemical constituents of *Achrostrichum aureum* Linn. Ayur, Hogg and Soper (1964) studied the nature of various alkaloids and investigated the alkaloids present in the *Lycopodium* species. Bohm (1968) investigated various phenolic compounds in ferns and proved indirect existence of 2,6-dihydroxy-4-4' dimethoxychalcone in *Pityogramma calomelanos.* Rangaswami and Iyer (1967, 69) have done chemical examination of *Adiantum vesnutum* Don. and also of *Chelienthes farinosa* flavonoids. Bohm (1968) have also studied the phenolic compounds in ferns and he examined some ferns for the presence of caffeic acid derivatives in it. Lahdesmaki (1968) isolated three amino acids from the leaves of *Salvinia natans* and *Azolla filiculoides* grown in light and dark. Bhabbie, Tewari and George (1972) studied the chemistry of *Actineopteris radiata* Link. Davidonis, and Ruddat (1973) studied the allelopathic compounds *Thelyterin A* and *B* in fern

Thelypteris normalis. Wang, Pamukeu and Bryan (1973) isolated fumaric acid, succinic acid, isoquercitrin and iliroside from the fern *Pteridium aquilinum.*

Chakravorti and Debnath (1974) studied the chemical constituents of leaves of *Marsilea minuta.* Nambudiry and Rao (1974) studied the terpenoids and synthesize Pterosin E, a sesquiterpenoid from Brakhen. Taneja and Tiwari (1974) studied the chemical constituents of *Actineopteris radiata.* Saleh (1975) studied the glycosidic nature of flavonoids isolated from the *Equisetum.* Singh, Rao and Hardikar (1975) studied the chemical constituents of *Adiantum caudatum.* Cooper (1976, 1977 and 1980) studied the chemotaxonomy and phytochemical ecology of Bracken. He also studied the role of flavonoids and related compounds in fern systematics. He gave phenolic chemotaxonomy and phytogeography of *Adiantum.* Gupta, Khanna and Sharma (1976) studied the chemical components of *Asplenium laciniatum.* Patil and Rao (1976) isolated the amino acids of the species of *Lastrea tenericaulis* Bedd. Wallace and Markham (1978) studied the flavonoids of primitive ferns like *Stromatopteris, Gleichinia, Hymenophyllum* and *Cordiomanes.* Lal (1979) studied the phenolic constituents of the therapeutical fern *Asplenium trichomanes* Linn. Hota and Patnaik (1982) investigated the distribution of amino acids in *Marsilea quadrifolia* Linn. Takahashi *et al.* (1984) identified even GA36 in the fern, *Psilotum nudum* that is a milestone in the field of phytochemistry of ferns. Kaur *et al.* (1986) have done a comparative investigation of amino acids and fern proline of some Rajasthan ferns. Sultane, llyas and Shaida (1986) investigated the chemistry of *Acrostichum aureum* Linn. Gaitonde and Garayelou (1988) isolated the flavonoids from the roots of *Lycopodium flixuosum.* Rathore and Sharma (1989) have studied the phytochemistry of Rajasthan pteridophytes, and studied the phenols in relation to stress. Maharan (1990) investigated the chemical composition and extracts of fronds of *Adiantum capillus-veneris* L. Umikalsom and Harborne (1991) studied the flavonol o-glycoside from the pinnae of *Asplenium marinum* Linn. Rathore and Sharma (1992) studied the phytochemistry of *Isoetes* L. found in Rajasthan. Sharma and Sharma (1992) also studied the flavanoids and phytochemistry of ferns of Rajasthan. Gopalkrishan, Rama and Suganthi (1993) studied the phytochemical nature of tree ferns of Western Ghats. Basil (1997) proved that the induction of antibacterial activity takes place by α-D-oligogalactacuronides in *Nephrolepis* sps. which is a pteridophyte. Singh (1999) studied the potential medicinal pteridophytes of India and their chemical constituents.

General Characteristics of Pteridophytes

Adiantum capillus-veneris L. (Adiantaceae)

Local Names: 'Hansraj' or 'Hansapadi'

Locality: Gwarparanath, Menal, Mt. Abu, Goramghat, Alwar.

Plants are found in sandy alluvial soil deposited in rock-crevices near waterfalls or under the moist and shady places on the humus rich soil. Rhizome slender, creeping to suberrect 10-30 cm long bearing fronds, roots and narrow lanceolate scales. Fronds are bipinnate, stipe slender, shining black; rachis slightly flexious and bears 5-6 secondary branches on either side, pinnae shortly stalked ending into a large terminal stalked pinnule. The lateral ones are shortly stalked more or less rhomboidal in shape, margin smooth or deeply lobed, texture thin, veins pellucid and distinct on both the surfaces, sterile pinnules may have rounded or finely toothed margins. Soral flaps infolded, semicircular to transversely elongate, 2-3 mm broad situated at the epices of lobed fertile pinnules. Sporangia leptosporangiate with tetrahedral, triangular, smooth walled spores each bearing a distinct triradiate mark.

The leaves are used as febrifuge cure; cough medicine and bronchial disorders (Chopra *et al.*, 1956; Bulletin N.B.G. 1959; Encl. I.B. and H. 1958; Dixit, 1975).

Adiantum incisum forsk. (Adiantaceae)

Local Name: 'Hansraj'

Locality: Common throughout the Aravalli Hills.

The rhizome is small and vertical, covered with numerous fibrous roots and scales. Fronds are pinnate 40 to 65 cm in size; each having a bud in its apical region, which serves the purpose of vegetative propagation, that is why this fern is called as 'Walking fern' The pinnae are opposite or alternate and placed separately. The stipe is dark brown, 0.9 to 1.5 mm in diameter and is densely covered with palae or hairs. The pinnae are 7 × 21 mm in size, sessile and are attached by their pointed bases. The lower margin is straight, while the upper is irregularly incised into truncate segments. The segments are of different sizes and the incision reach from 1/2 to 2/3 down in the lamina. Venation is conspicuous on the upper surface of the laminae. In the fertile frond each pinnae bears a number of marginal separate sori, which are generally long in shape. The sori are partly covered by a laminar–flap or indusium that has only few hairs on it. Large number of sporangia is produced in each sorus. Approximately 32 spores are produced per sporangium.

It is used in cough, diabetes and skin diseases. (Sharma and Vyas, 1985). In Mt. Abu area the Tribal race 'Bheel' uses the juice of leaves in skin diseases. In Goramghat area the powder of leaves is mixed with butter and is used for controlling the internal burning of the body. The Garasia Tribe people mix the dry leaves with tobacco and smoke to curb the internal burning of the body.

Adiantum lunulatum Brum. (Adiantaceae)

Local Name: 'Hansraj'

Locality: Very common at Mt. Abu, also found at Gwaparanath (Kota), Ajmer and Menal (Chittore).

The rhizome is ascending, small, 8 to 22 mm in size, densely covered with fibrous roots, scales and leaf bases. Frond pinnate, 16.5 to 42 cm in length, stipe shining dark brown with alternate, sub-opposite or opposite pinnae. The young leaves show circinate venation and are densely covered with palae. In a mature frond the multicellular palae occur only at the base. The pillae are stalked, stalk dark brown, naked, 1.5 to 2.0 mm thick. The size of stalk reduces gradually towards the apex *i.e.* it is quite big in the lower pinnae, while the upper pinnae are almost sessile. The pinnae are more or less lanceolate, 0.8-3.2 cm in length, having lower margins curved and lobed, irregularly or regularly. The lobes are round or truncate. The lamina is soft, naked and glossy. In the fertile frond the outer margin of pinnae is smooth bearing an almost continuous sorus (tetraploid plants) or it is broken into 3-5 groups (triploid plants). The sori are longish and protected by the marginal flap or indusium. Each sorus possesses large number of sporangia having vertical annulus, made up of 13 to 17 cells. 32 to 48 spores are produced per sporangium.

Ayurvedic Vaidyas describe the plant as pungent, alexiteric and indigestible. They consider it useful in dysentery, diseases of the blood, ulcers, and erysipelas. (Cains, 1935 and Chopra *et al.* 1956). Sporophylls are used in leprosy and erysipelas. Nadkarni (1982) and Chopra, Dixit (1975) described the use of *A. lunulatum* in leprosy, fever and erysipelas.

Sharma and Vyas (1985) described that the local people in the Aravalli Hills use the decoction of this plant in cough, 'Asthma' and fever. The paste of leaves is used in leprosy and erysipelas. It is also used to overcome 'Hair falling' by putting paste of its leave on head for an hour or so before taking bath, for a 'Fortnight'.

In Mt. Abu area, the 'Bheel' uses this plant for urine diseases and in bleeding from nose. In former the extract of leaves is taken orally and the paste of leaves is applied on the lower portion of stomach, for clear and early release of urine. The leaves extract is put in drops into the nose to stop bleeding, during summer months.

Actineopteris radiata (Swartz.) Link (Actineopteridiaceae)

Local Names: Morpankhi, 'Sanjeevani', 'Jahreela Podha', 'Baliyar', 'Morsetti', 'Morthotha'.

Locality: Throughout the Aravalli Hills between 400-1000 meter MSL (Ajmer, Alwar, Jello, Dabla, Mt. Abu, Sirohi, Bundi, Udaipur, Goramghat, Sundhamata (Jalore) etc.

The plants are 8-25 cm high, rooting in the crevices of rocks or in between the joints of brick walls in moist and shady places. The rhizome is oblique to horizontal, 1.5 to 2.0 cm in length, densely covered with wire roots, palae and leaf bases. The palae are multicellular, each having a terminal gland. The young leaves show circinate venation but the lamina becomes flat at an early stage of development. The stipe is 3 to 20 cm long provided with scales on its entire length in the young leaf, while they are sparsely placed in amature frond. The lamina is divided into two groups, which dichotomize further into segments. The laminae are stiff and rough to touch. The sporangia are sub-marginal on an inter-marginal vein covering almost the entire abaxial surface of segment.

Chopra *et al.* (1956) and Dixit (1975) described the plant styptic and antihelminthic. Sharma and Vyas (1985) described its use in bronchitis and gynecological disorders. In Goramghat, Kumbhalgarh, and Parshuram Mahadev area the ash of the leaves is taken with honey, 2-3 times a day to get relief from bronchitis. Similarly, the paste of 5-6 leaves mixed with fresh cow milk (nearly 200 ml) is taken for a week or so, to over come irregularity in menstrual period. The ash (approximately 2-3 gms.) of the plant mixed with fresh cow milk (200 ml) is given to a lady for a fortnight after menses period that wishes to have an issue. On the other hand the paste of 8-10 leaves mixed with thin curd (nearly 250 ml) is given for birth control. Decoction of leaves is also used in tuberculosis in the Mt. Abu area by Bheels.

Araiostegia pseudocystopteris Copel. (Dennstaedtiaceae)

Locality: Mt. Abu

Terrestrial rhizome creeping clothed with brown scales. Stipe glabrous, jointed to the rhizome. Fronds tripinnate pinnae much dissected into small rhomboidal segments clefted into two oblique elliptic lobes. Sori globose to subglobose, near the sinus. Indusia suborbicular, attached only at base, sides free. Sporangia long stalked, annulus oblique, 8-13 celled. Spores bean-shaped hyaline, exine smooth or slightly grooved.

The Tribal people at Mt. Abu use the decoction of fronds as vermifuge.

Asplenium pumilum var *hymenophylloides* (Aspleniaceae)

Locality: Mt. Abu

Rhizome short, ascending, scales calthrate, hairs usually present. Fronds tufted, stipe not articulate to rhizome. Lamina deltoid, lowest pair largest, pinnae pubescent or ciliate on both surface, textures extremely thin. Sori elongate along veins, fully covered with indusium when young, opens towards costules.

Sporangia long stalked, annulus incomplete, vertical, spores bilateral.

Athyrium pectinatum (Well.) Presl. (Athyriaceae)

Local Name: 'Jaributi'

Locality: Mt. Abu

Rhizome creeping and branched; scales brown, lanceolate and up to 7 to 15 mm in length; stipes fragile, straw coloured, 10-35 cm long, lamina variable, broadly lanceolate to sub deltoid with acuminate apex, decompoundly pinnate and finely dissected, pinnae stalked, distantly placed, 6-15 × 2-4.5 cm ascending with slender, naked, greenish rachides, pinnules, up to 15 × 6 mm, sub deltoid, cut down into ultimate oblong, narrow segments with dentate margin, secondary rachides minutely pubescent, veins forked, sori minute, indusium thin, membranous, spores dark brown, Perrine, broad with wrinkles and anatomizes on the surface.

The plant is common in Mt. Abu area and is frequently used by the Bheels for medicinal purpose. The young leaves are used as vegetable. The rhizome is considered as a strong antihelminthic.

Chelienthes albomarginata Clarke (Sinopterideceae)

Local Names: 'Dodhari', 'Nanha'

Locality: Common throughout the Aravalli Hills

Rhizome short, covered with fibrous roots and ramenta. Scales brown, lanceolate with needle-like apex and translucent margins. Stipe brown/black, slender, erect, 5–10 cm long covered with scales. Lamina lanceolate-deltoid, 10–25 cm long, at the base, unipinnate with deeply pinnatified pinnae, hairy and scaly, basal pair of pinnae more developed than the others, scaly, white powdery on ventral surface of lamina, dark green above. Sori marginal, confluent, indusium greenish-brown, margin lacerates, Sporangia leptosporangiate, spores with wrinkled/smooth exine.

The plant is used by Tribal people, Bheel, Meena, Garasia and Saharia, etc. as a general tonic for children and weak people. The extract of the plants (leaves) mixed with honey is taken after meal by the Tribal people suffering from tuberculosis.

Cyclosorus dentatus Holtum (Thelypteridaceae)

Locality: Mt. Abu

Rhizome short and creeping. Stipe vary variable in length, hairy, scaly at base, scales narrowly lanceolate. Lamina up to 90 cm long, pinnae 15–25 pairs, lower 2–4 pairs gradually reduced up to 4–5 cm long, distantly placed, auricled at macroscopic base, auricle lobed. Largest pinnae 8-10 × 1.5-2 cm long, often larger, lobes more than half way to costate, slightly oblique with rounded apex.Veins 8–9 pairs, lowest pair anatomizing with excurrent vein to sinus, immediate next pair to the side of sinus membrane. Both surface have costate and costules hairy, sori medial, Indusia hairy.

The 'Bheel' people used young circinate leaves as vegetables at Mt. Abu.

Dryopteris cochleata (Don) C. Chr. (Dryopteridaceae)

Local Name: 'Jaributi'

Locality: Mt. Abu

Rhizome woody, stout, horizontal or ascending, thickly covered with leaf bases and brown scales. A tufted large fern with generally dimorphic fronds. Barren fronds pinnate or sub-bipinnate and approximately a meter in length; pinnae 12-18 pairs, close, lowest often 30 × 8 cm, oblong, lanceolate pinnules lobed or not, denticulate or serrate, and glabrous. Fertile fronds narrowly lanceolate and smaller than vegetative fronds 2-pinnate with pinnules much contracted thickly covered with sori.

The young leaves are used as vegetable. Tribal people also use in eczema and as antihelminthic plant.

Equisetum ramosissimum: sub. sp. *ramosissimum* Desf.

Local Names: Jangli bans or Ghas bans.

Locality: Ajmer, Kota, Jaipur, Mt. Abu etc.

Rhizome creeping, slender, stem erect, 3.0 cm. to 1.2 meter high, simple and slender or irregularly branched in whorls, vegetative and fertile branches identical. Stem ribbed and jointed. Ribs 8-10; prominent. Sheaths 5-8 mm long, teeth prominent 8-10, black pointed, lanceolate acuminate with a deltoid base, deciduous terminal branch fertile. Spikes (cones) oblong, 1.2–2.5 cm long and 1-6 mm in diameter. Large and small specimens appear very different superficially.

The plant is administered as cooling medicine and is given for gonorrhea (Chopra *et al.*, 1956).

Hypodematium crenatum (Forsk.) Kuhn (Hypodematiaceae)

Local Name: 'Jaributi'

Locality: Mt. Abu, Gwaparnath, Chittorgarh, Parshuram Hills.

Rhizome creeping, dorsiventral and densely clothed with light brown palae or thin golden yellow scales, pinnules and indusium also setose throughout with white setae. Sori abaxial, reniform, greenish white when young, dark brown on maturation. Spores trilete with distinct triradiate markings.

Sharma and Vyas (1985) described that the plant is given to ladies for conception; the paste or powder of leaves along with fresh cow milk is taken, five days after the menstrual period for about a week (Sharma and Vyas, 1985). The plant is also used for getting relief from insect bite in Mt. Abu area. The woodcutters at the point of bite or injury apply paste of the frond.

Isoetes rajasthanensis Gena and Bhardwaja (Isoetaceae)

Local Name: 'Ghas Phus'

Locality: Mt. Abu

The plants are medium, 5 to 16 cm in length. The rhizomorph is 2-3 lobed; rarely four lobed structure having 6 to 12 leaves per plant. The roots arise in regular sequence from the lower portion of the rhizomorph. Velum covers 1/3 portion of sporangium, Microsporangia rare. Microspores monolete, megaspores trilete and spinose.

Marsilea minuta L. (Marsileaceae)

Local Name: 'Choupatha'

Locality: Alwar, Ajmer Mt. Abu, Sirohi etc

Plant aquatic/amphibious, rhizome, a runner with distinction of nodes and internodes. From the nodes arise roots, leaves and sporocarps. Leaf with large petiole terminating into four leaflets (two close pairs of opposite and decussate leaflets). Leaflets obovate with smooth or crenate margins, veins many, frequently dichotomised. Sporocarp stalked, bean shaped rounded.

Plants are used in cough, spastic condition of leg muscles etc. and also in sedatism and insomnia (Ray and Gupta, 1965; Dixit, 1975). 'Garasia' and 'Bheel' cook the leaves as vegetable. The decoction of leaves along with ginger is used to cure cough and bronchitis in many villages of Rajasthan.

Ophioglossum nudicaule L. (Ophioglossaceae)

Local Name: 'Ghas phus'

Locality: Mt. Abu, Menal, etc.

Plants small, 2.9 to 6.2 cm tall, rhizomes globose, rarely thick cylindrical giving rise to fibrous roots. Aerial parts with small common stalk, sporophyll ovate or eliptical, lanceolate, apex acute, or obtuse, base cordate, alternate, or truncate, texture fleshy, venation lax to dense with free veinlets. Fertile stalk 2.2 to 4.2 cm long, spike small 3 to 14 mm in length.

Tectaria coadunata (J. Smith) C. Chr. *(T. macrodonta)_(Aspediaceae)*

Local Name: 'Jadibuti'

Locality: Mt. Abu

Rhizome large ascending or horizontal covered with roots, scales and leaves. The fronds are large and often 0.5–1.0 meter long and somewhat deltoid in outline, pinnatified or distinctly pinnate below with the pinnae deeply pinnatified, rarely bipinnate or more compound. Rachis surface glabrous, stipes tufted, chestnut brown with scaly base. Main vein fairly distinct to the margin, others anatomizing with often free included veinlets. Sori large in two rows between the main veins, on the netted veins or at the apex of free veinlets with a reniform or usually peltate indusium. Sporangia typical leptosporangiate with vertical annulus, 48-64 monolete spores are produced per sporangium.

According to Mehra and Khullar (1974) the correct nomenclature of *T. macrodonta* is *T. coadunata*.

The plant occurs frequently at Mt. Abu so the 'Bheel' uses it frequently as medicinal plant. The leaves mixed with honey or decoction of leaves is given to asthma and bronchitis patients. Woodcutters use the paste of leaves on the place of irritation caused by stings of honeybee, centipeds etc. (Sharma and Vyas, 1985).

References

Ayer, W.A., Hogg, A.N. and Sper, A.C., 1964. *Lycopodium* alkaloids VI. The nature of alkaloids X 9. *Canadian J. Chem.*, 42: 949–951.

Banerjee, R.D. and Sen, S.P., 1980. Antibiotic activities of pteridophytes. *Economic Botany*, 34(3): 284–98.

Basil, A., Spagnulo, V., Castaldo S., Sorrentino, C., Lavitola, A. and Castaldo, C.R., 1997. Induction of antibacterial activity by α-D-oligogalacturonoides in *Nephrolepis* sps. (Pteridophyta). *International J. of Antimicrobial Agents*, 8(2): 131–134.

Bhabbie, S.H., Tewari, P. and George, C.X., 1972. Chemical analysis of *Actiniopteris radiata* (SW.) Link. *Current Science*, 41(2): 788.

Bhardwaja, T.N. and Garg, A., 1984. The antifertility effect of an Australian species of the aquatic fern *Marsilea* L. *Indian Fern J.*, 1: 75–82.

Bohm, B.A., 1968. Phenolic compounds in ferns-III: An examination of some ferns for the presence of caffeic acid derivatives. *Phytochemistry*, 7: 1825–1830.

Cain, J.F., 1935. The medicinal and poisonous ferns of India. *J. Bombay Nat. Hist. Soc.*, 38: 341–361.

Chakravarti, D. and Debnath, N.B., 1974. Chemical constituents of leaves of *Marsilea minuta*. *J. Indian Chem. Soc.*, 51: 260–265.

Chatterjee, A., Dutta, C.P., Choudhary, B., Dey, P.K., Dey, C.D., Chatterjee, C. and Mukherjee, S.R., 1963. The chemistry and pharmacology of *Marsilea*: A sedative and anticonvulsant principle isolated from *Marsilea minuta* L and *M. rajasthanensis* Gupta. *Sci. and Cult.*, 29: 619–602.

Choudhary, A.K., Ali, M.S. and Khan, M.O.F., 1997. Antimicrobial activity of *Ipomoea fistulosa* extracts. *Fitoterapia*, 68: 4; 380–397: 9th.

Cooper Driver, G., 1976. Chemotaxonomy and phytochemical ecology of Bracken fern. *Bot. J. Linn. Soc.*, 73: 35–46.

Cooper Driver, G., 1980. Role of flavonoids and related compounds in fern systematics. *Bull. Torrey, Bot. Club,* 107: 116–1270.

Cooper Driver, G., 1997. Phenolic chemotaxonomy and phytogeography of *Adiantum. Bot. J. Linn. Soc.*, 73: 35–46.

Davidonis, G.H. and Ruddat, M., 1973. Allelopathic compounds Thelyterin A and B in fern *Thelypteris normalis*. *Planta*, 111: 23–32.

Deniel, M., 1991. *Methods in Plant Chemistry and Economic Botany*. Kalyani publishers, New Delhi.

Devi, S., Jamil, Z. and Verma, R.C., 1980. Allergic manifestations of fern spores elliated in humans. *J. Polynol.*, 16: 115–123.

Devi, S., Jamil, Z. and Verma, R.C., 1980. Fern spores as potential aeroallergens. In: *Contemporary Ferns in Plants Sciences*, (Ed.) S.C. Verma. Kalyani Publ, New Delhi, pp. 221.

Dhar, M.L., Dhar, M.M., Dhaman, B.N., Mehrotra, B.N. and Roy, C., 1968. Screening various Indian ferns for biological activity. *Indian J. Exptl. Biol.*, 6: 232–247.

Dixit, R.D., 1974. Fern: A much neglected group of medicinal plants. I. *J. Res. Indian Med.*, 9: 74–90.

Dixit, R.D., 1975. Fern: A much neglected group of medicinal plants. III. *J. Res. Indian Med.*, 19: 309–314.

Dixit, R.D. and Bhatt, G.K., 1975. Fern: A much neglected group of medicinal plants. II. *J. Res. Indian Med.*, 10: 68–76.

Dixit, R.D., Das A. and Kar, B.D., 1978. Studies in ethnobotany III of some less known edible, economic and medicinal ferns of Darjeeling District, West Bengal, Nagarjun, 21: 1–4.

Ganesan, T., 1993. Effect of leaf extracts of some pteridophytes on *Drachslera oryzae* conidial germination. *Geobios*, 20: 262–263.

Gehlot, Dushyant and Bohra, A., 1995. Antibacterial activities of leaf extracts of some ferns from Pachmarhi Hills. In: *Abst. Proc. of National Symposium on Researches in Pteridology*. October 5–7 at J. N.V. University, pp. 118.

Gehlot, Dushyant, Bohra, S.P. and Bohra, A., 1995. Screening of some ferns plants for antifungal activity. In: *National Symposium on Researches in Pteridophytes, Abst. of Papers*, October 5–7.

Gehlot, Dushyant, Gupta, V.B. and Bohra, A., 1995. Antibacterial activities of leaf extracts of some ferns from Pachmarhi Hills. In: *National Symposium on Researches in Pteridophytes, Abst. of Papers*, October 5–7.

Gopalkrishan, S., Rama, V. and Suganthi, Angelin, 1993. Phytochemical studies on tree ferns of Western Ghats. *Indian Fern J.*, 10: 206–213.

Gupta, R.B., Khanna, R.N. and Sharma, N.N., 1976. Chemical components of *Asplenium laciniatum Current Science*, 45(2): 44–46.

Hansraj, H., 1996. Medicinal value of *Adiantum. Indian Drugs*, 34: 36.

Horsely, S.B., 1977. Allelopathic inhibition of black cherry by ferns, grass, golden rot and asters. *Cand. J. For. Res.*, 7: 205–216.

Hota, G. and Patnaik, R.K., 1982. Distribution of amino acids in *Marsilea quadrifolia* Linn. *Proc. Indian Sci. Congr.*, Part 3, 69: 66.

Kapur, S.K. and Sarin, Y.K., 1977. Useful medicinal ferns from Jammu and Kashmir. *Indian Drugs*, 14(7): 136–140.

Kaur, R., Yadav, B.L. and Bhardwaja, T.N., 1986. A comparative investigation of amino acids and fern proline of some Rajasthan ferns. *Bionature*, 6: 42–44.

Kaushik, P. and Dhiman, A K., 1995. Common medicinal pteriodophytes. *Indian Fern Journal*, 12(1–2): 139–145.

Kaushik, P., 1988. *Indigenous Medicinal Plants including Microbes and Fungi*. Today and Tomorrow's Printers and Publishers, New Delhi.

Kshirsagar, M.K. and Mehta, A.R., 1972. Survey of ferns in Gujarat State (India) for presence of antibacterial substances of ferns. *Planta Med.*, 22(4): 386–390.

Kumar, A. and Roy, S.K., 1972–73. Some medicinal ferns from Netharhar hills (Bihar). *J. Sci. Res., B.H.U.*, 23: 139–142.

Lal, S.D., 1979. Phenolic constituents of the therapeutical fern *Asplenium trichomanes* Linn. *Sci. and Cult.*, 45(11): 452–453.

Lall, G., Kapoor, J.N. and Munjal, R.L., 1964. Some ferns-inhibiting *Cercospora species* from India. *Ind. Phytopatho.*, 17(2): 181.

Mahran, G.H., El-AIfy, T.M., Taha, K.F. and El Tantawy, M., 1990. Chemical composition and antimicrobial activity of volatile oil and extracts of fronds of *Adiantum-Capillus veneris* L. *Bulletin of Faculty of Agri. Univ. of Cairo*, 41(3): 555–572.

Nayar, B.K., 1958. Medicinal ferns of India. *Bull. Nat Bot. Garden*, 29: 1–36.

Nayar, B.K., 1958. An enumeration of the medicinally used pteridophytes of India. *Med. Pl. Symp. Proc. (CSIR)*, p. 6–8.

Pandey, A.K. and Bhargava, K.S., 1980. Antiviral activity of crude extracts of some pteridophytes. *Indian. Ferns. J.*, 3: 132–133.

Parihar, P. and Bohra, A., 2001. Effect of some pteridophytic plants found in Rajasthan on the growth of *Escherichia coli*. In: *Abst. Proc. of National Conference on Emerging Areas in Plant Sciences*, October 5–6, p.16.

Parihar, P. and Bohra, A., 2002. Antibacterial efficacy of some pteridophytic plants (found in Rajasthan) against human pathogenic bacteria–*Staphylococcus aureus*.

In: *Abs. Proc. of National Seminar on Role of Antimicrobials for Sustainable Horticulture.* January 20.

Parihar, P. and Bohra, A. Anticandidal activity of some pteridophytic plant part extracts. *Journal of Microbial World,* 4(2) (In press).

Parihar, P. and Bohra, A., 2000. A comparative study of filter paper disc method and microbial bioassay method for the evaluation of antimicrobial activity by plants against pathogenic organisms. *Journal of Eco-physiology,* 3(1–2): 23–25.

Parihar, P. and Bohra, A., 2002. Antifungal efficacy of various pteridophytic plant parts: A study *in vitro. Advances in Plant Sciences,* 15(1): 35–38.

Parihar, P. and Bohra, A., 2002. Effect of some pteridophytic plant part extracts on human pathogenic bacteria–*Salmonella typhi. Indian Fern Journal,* 19 (In press).

Parihar, P. and Bohra, A., 2002. Screening of some ferns for their antimicrobial activity against *salimonella typhi. Advances in Plant Sciences,* 15(2): 365–367.

Parihar, P. and Bohra, A., 2002. Antibacterial efficacy of various pteridophytic plants extract against *Escherichia coli:* A study *in vitro. Ecobios,* 1(1): 7–9.

Parihar, P., Daswani, L., Bohra, A. and Bohra, S.P., 2002. Antibacterial activity of *Ricia arravelliensis* (Pande et udar) and *Plagiochasma appendiculatum (Lahm et Lindenb). Bioscience Research Bulletin,* 18(1): 61–63.

Paulo, A., Duarte, A. and Gomes, E.T., 1994. *In vitro* antibacterial screening of *Cryptolepis sangunilinta* alkaloids. *Journal of Ethnopharmacology,* 44(2): 127–130.

Paulo, Daurte and Gomes, 1997. Chemical and antimicrobial studies on *Cryoptolepis obtusa* leaves. *Fitoterapia,* 68(6): 558–559.

Puri, G.S. and Arora, R.K., 1961. Some medicinal ferns from Western India. *Indian Forester,* 87: 179–183.

Puri, G.S., 1971. Phytochemical survey of some plants for steroids, saponion. alkaloid and tannins. *Indian Drugs,* 8: 7–10.

Rangaswami, S. and Iyer, R.T., 1967. Chemical examination of *Adiantum vesnutum* Don. *Curr. Sci.,* 36: 88–89.

Rangaswami, S. and Iyer, R.T., 1969. Flavonoids of *Chelianthes farinosa. Indian. J. Chem.,* 7: 526.

Rajappan, K., Mariappan, V. and Kareem, A. Abdul, 1997. Effect of dried leaf extract of *Ipomea* on rice sheath rot pathogen and beneficial micro-organism. *Indian Phytopathology,* 50(1): 329.

Rathore, D. and Sharma, B.D., 1989. Phytochemistry of Rajasthan pteridophytes: Study of 20 phenols in relation to stress. *Indian Fern. J.,* 6: 244–246.

Rathore, D. and Sharma, B.D., 1990. Phytochemistry of Rajasthan ferns in relation to stress. *Indian Fern. J.,* 7: 184–187.

Rathore, D. and Sharma, B.D., 1992. Pteridophytes of Rajasthan: Phytochemistry of *Isoetes* L. *Indian Fern J.,* 9: 121–122.

Raymundo, A.K., Tan, B.C. and Asuncion, A.C., 1989. Antimicrobial activities of some Phillipine cryptogams Phillipine. *Journal of Science*, 118: 59–65.

Saleh, N.A.M., 1975. Glycosidic nature of *Equisetum* flavonoid. Phytochemistry, 14(1): 286.

San-Fransisco, M. and Cooper Driver, G., 1984. Antimicrobial activity of phenolic acid in *Pteridium aquilinum*. *American Fern Journal*, 74(3): 87–96.

Sen, S. and Nandi, P., 1951. Antibiotic from the pteridophytes. *Sci. and Cult.* 16: 328–329.

Sharma, A. and Sharma, B.D., 1992. Phytochemistry of Rajasthan Ferns: A study of flavonoids. *Indian Fern. J.*, 9: 83–86.

Singh, H.B., 1999. Potential medicinal pteridophytes of India and their chemical constituents. *Journal of Economic and Taxonomic Botany*, 23(1): 63–78.

Singh, S., Rao, M.N.A. and Hardikar, S.G., 1975. Chemical constituents of *Adiantum caudatum*. *Indian J. Pharm.*, 37: 64–65.

Singh, S.P. and Roy, S.K., 1986. Some medicinal ferns from the Mirzapur (Hathimala) forests. *Bull Medico-Ethnobot. Res.*, 7(3 and 4): 185–187.

Singh, V.P., 1973. Some medicinal ferns of Sikkim Himalayas. *J. Res. Indian Med.*, 8: 71–73.

Srivastava, S.L. and Kediyal, V.K., 1984. Effect of fern extracts on conidial germination and germ tube growth of two pathogenic fungi. *Indian Phytopathology*, 37(3): 561–563.

Srivastava. S N., Bbhakuni, D.S. and Sharma, V.N., 1963. Chemical constituents of *Achrosstrichum aureum* Linn. *Indian J. Chem.*, 1: 499.

Taneja, S.C. and Tiwari, H.P., 1974. Chemical constituents of *Actiniopteris radiata*. *Curr. Sci.*, 43: 749–750.

Vyas, M.S. and Sharma, B.D., 1988. Ethnobotanical importance of ferns of Rajasthan. In: *Proc. Indigenous Medicinal Plants*, Gurukul Kangri, Today and Tomorrows, New Delhi, India, p. 61–66.

Chapter 6

Taxonomical and Ethnomycological Observations of Edible Mushrooms of Arunachal Pradesh

Rishikesh Mishra[1] and Anil Kumar Sinha[2]

[1]*Department of Botany, D.N. Government College,*
Rajiv Gandhi University, Itanagar, Arunachal Pradesh
[2]*Department of Botany, K.C.T.C. College,*
Raxaul – 845 305, Constituent Unit of B.R.A. Bihar University,
Muzaffarpur, Bihar

Introduction

Arunachal Pradesh is a new name of the territory which was included under the erstwhile North East Frontiers Agency (N.E.F.A.). It is spread along the south face of the Eastern Himalayas immediately below the water divide between Tibet on one hand and India on the other and the water divide between Burma and India along the crest of the Patkoi hills coming round to the South East and South of the Eastern most trip of the Brahmputra Valley. It is situated between latitudes 26°38′N and 29°30′N and Longitudes 91°30′E and 97°30′E covering an area of 83,578 sq. km. with massive forest cover (about 60 per cent of the land) at an altitude ranging from 170 to 5,000 m resulting in varying climatic zones of hot valleys at the foot hill regions and the snow clad peaks at higher elevations.

A high degree of acidity is reported in soil which may be attributed to heavy rainfall. The soils are sandy to sandy loam. The soil in the foothills is alluvial in characters being either loamy or sandy or both mixed. In general, soils have rich layer of organic matter due to the rottening of the jungle leaves.

Arunachal Pradesh is included within the regions of heaviest rain fall in India. The incidence and condition of precipitation varies from one part to the other depending on great contrast in temperature between the low lying areas and regions of higher altitude; between sheltered valleys and exposed mountain slopes in the north. The river valleys nearer to and opening out to the plains of Assam receive the moisture laden monsoon clouds by a natural process of suction into great depths of the regions.

Arunachal Pradesh is very rich in mushroom flora on account of witnessing ideal climatic conditions for their luxurious growth. But unfortunately no systematic exploration of these mushrooms has been conducted so far to bring them on record. In view of the paucity of information prevailing on the mushroom flora of Arunachal Pradesh, a maiden attempt has been made by the authors to undertake extensive survey of Arunachal Pradesh for collection and identification of the mushrooms growing under different ecological situations.

Methodology

The methodology used for taxonomic studies, edibility and mycorrhizal associations, is as follows:

Collection

Mushrooms generally grow under humid conditions. The edible mushrooms in Arunachal Pradesh begin to appear immediately after the first showers in the later half of April and flourish well during June to September and occasionally even up to early October depending upon the frequency of rains.

Collections of edible mushrooms were done during the rainy seasons from 1995 to 1998. Places of collection along with its attitude are summarized in the Table 6.1.

The specimens were collected by picking up individually the basidiocarps by carefully digging them with the help of a sharp knife. Collections of fruiting bodies at all stages of development were attempted. Usually data of habitat temperature and humidity were recorded in the field itself. Each collection was wrapped in a paper bag and brought carefully to the laboratory for further studies. The necessary tools used for the collections of mushrooms were a basket, a few small containers, glass or plastic tubes, brown paper bags for larger fungi, a hand lens and a pen knife. Ecological notes were also recorded.

Preservation

The specimens were preserved dry as well as wet. Wet preservations were done in alcohol/formaline in glass jars following the method of Ainsworth (1971). The mushrooms were dried in an oven and were poisoned with mercuric chloride and then kept in paper bags.

**Table 6.1: Various Sites Explored Periodically for
Collection of Edible Mushrooms Along with Altitude**

Sl.No.	Collection Sites	District of Arunachal Pradesh	Altitude in Meters
1.	Pasighat	East Siang	155.00
2.	Mebo	East Siang	346.00
3.	Nari	East Siang	180.00
4.	Ruksin	East Siang	133.00
5.	Along	West Siang	350.00
6.	Basar	West Siang	650.00
7.	Yingkiong	Upper Siang	428.00
8.	Pangin	Upper Siang	455.00
9.	Boleng	Upper Siang	238.00
10.	Meriyang	Upper Siang	1067.00
11.	Jenging	Upper Siang	375.00
12.	Gelling	Upper Siang	1050.00
13.	Bilat	East Siang	167.00
14.	Ziro	L. Subansiri	1560.00
15.	Itanagar	Papumpare	550.00
16.	Diomukh	Papumpare	450.00
17.	Roing	Dibang Valley	390.00
18.	Tezu	Lohit	210.00

Identifications

The identification of mushroom species require recording of a number of characters in the field *e.g.* spore, colour, habitat, forest type, pileus and gill structure, colour change induced after cutting, bruising and exposure in different parts of hymanophore, macroscopic characters of stipe etc. These characteristics were carefully recorded in the field.

Methods and terminology used for the description of mushrooms were as given by Henderson *et al.* (1968), Malloch (1971), Singer (1975), Largent (1977a), Soothll and Fairhurst (1978), Purkayastha and Chandra (1985) and Lakhanpal (1995, 1996).

Macrochemical colour reactions were performed on mature sporocarps. A sporocarp was split vertically dividing it into two halves, slices of exposed flesh approximately 2 mm thick and 5-7 mm deep were cut from the exposed surface avoiding the margin of cap. The following reagents were used for macrochemical reactions.

1. Melzers Reagent (Largent, 1977b)
2. Fresh 10 per cent solution of $FeSO_4$

Three to four drops of reagent were dropped on these tissue slices and the colour change, if any after 10–30 minutes was noted.

The edibility of the species was ascertained by enquiries made from the local inhabitants, the residents of the neighboring villages of the collection sites. They were asked about the edibility of the collections and the methods of preparation and consumption. Local names and recipes were also noted. The edibility of the species was also confirmed from literature.

Taxonomical Enumerations

Altogether 38 taxa of edible mushrooms have been collected in the present endeavors. Taxonomical enumeration of the collected taxa along with their ecological details may be read under the respective heads as follows:

Agaricus arvensis Schaeff ex. Seer.

Pileus 8.0–16 µm in diam, convex semi-ovate or flattened, colour-creamy white, soft, surface smooth, sometimes flat, scales present gills-crowded free, grayish pink, finally dark brown, stipe-central, cylindrical 6.0–12.0 cm long, 1.0–3.0 cm thick, smooth stout, bulbous, basidia; broad and squat 12-21 × 7-8 cm volva-absent, ring; superior, appearing double, basidiospores; dark purple brown, elliptical and smooth, 6.5-8 × 4.5–5.5 µm, spore print purple brown.

Habitat

It was collected from Pasighat and Yinkiong in the month of April 1995. It was found in pastures, grasslands.

Habit

Solitary (Figure 6.3) Atm. Temperature 23°C, Relative Humidity (R.H.)–88 per cent, Temperature of soil–26°C. The moisture of soil 72–82 per cent and soil pH 6.1–6.2.

Agaricus compestris Fies

Pileus 4.5–9.0 cm broad, hemispherical then convex and expanded, colour whitish, soft and fleshy; context 6–9 µm thick, surface smooth to slightly fibrillose, scaly, velar remains adhered to the margin gills crowded and free, white to pink, finally dark brown, stipe central, cylindrical or tapering downwards without volva with a ring, 5.0–8.0 cm long, flesh white, basidia 4 spored 24.0-26.0 × 7.0-9.0 µm, basidiospores dark brown, 7-8 × 3.5–5.5 µm.

Habitat

It was collected from Pasighat in the month of March, 1996. It was found growing in pastures or fields forming fairy rings.

Habit

Solitary, Atm. Temperature 24°C, RH–91 per cent, Soil temp–26°C, Soil moisture–78 per cent and Soil pH–6.2.

Agaricus subrufescens Peck

Pileus 9.5–11.5 cm, in diameter, plano-convex, convex finally broadly expanded surface smooth, white to reddish brown; fleshy soft, covered by reddish brown scales stipe stout and whitish, slightly swollen at the base, gills crowded and free, dark brown basidia 2–3 spored, 25.0-32.0 × 7.0–8.0 µm, basidiospores purple brown, elliptical 7-8.0 × 3.5–5 µm spore print brownish.

Habitat

It was collected from Pasighat in the month of March, 1995 and Sille in April 1997, in forests or woodlands on soil.

Habit

Generally in groups but solitary also; Atm. Temp–21°C, R.H.–90 per cent, Soil Temp–24°C, Soil moisture–81 per cent Soil pH–6.2.

Agaricus brunnescens Peck

Pileus 5.0–10 cm wide, convex when young, plano convex to flattened when old, whitish to pale brown, soft, surface smooth to finally scaly, fibrillose, striate, stipe central, uniform or slightly attenuated at the top, sometimes with bulbous base, 3.0–12.0 cm long, 1.0-1.8 cm thick, gills dull pink then dark brown, crowded and free, basidia usually bispored, clavate, basidiospores brown ellipsoidal, 6-8 × 5-6 µm, spore print sepia colored.

Habitat

It was collected from Balek village near Pasighat in the month of April, 1997. It was growing on manure heaps in gardens on soil rich in manure.

Habit

Mostly solitary, sometimes in groups, Atm. Temp–27°C, RH–91 per cent, Soil Temp–28°C, Soil moisture–82 per cent, Soil pH–6.4.

Agaricus sylvaticus Schaeff. ex. Secr.

Pileus 5.0–11.0 cm in diameter, convex or companulate when young, later expanded, surface whitish brown, brown or grey, soft, fleshy, fibrillose or with brownish or reddish brown oppressed scales, stipe central, uniform 1.0-2.5 cm thick, flesh white, turn reddish when cut gills free, reddish-brown, volva absent, spore print; cigar brown or brown; basidia; four spored; cheilocystidia clavate and thin walled basidiospores, purplish brown, 3.0–3.6 µm, globose to subglobose.

Habitat

It was collected from Pasighat in the month of July, 1995 and from Pangin in April, 1997. It was found on the ground in forests and in fields adjacent to woods.

Habit

Solitary or in group, Atm. Temp-22° C, R.H.-91 per cent, Soil Temp–24°C, Soil moisture–81 per cent, Soil pH–6.2.

Amanita caesarea (Fries) Quelet

Pileus; 8.0-17cm in diameter, ovoid or ovate, then convex, finally plane, surface; orange, reddish orange or yellowish orange; smooth, margin striate, fleshy soft, viscid in humid conditions, stipe yellow cylindrical and thick with bulbous base, enveloped by volva, yellow ring present, volva white large,12.7-19.5 cm long, annulus yellow or orange; gills crowded and free, yellow or orange flesh firm and white, yellowish under cap cuticle with an pleasant taste and odour, spore print; white, basidiospores; white to tinged yellowish, ellipsoid,10-14 × 6–12 µm. hyaline amyloid, thin walled.

Habitat

It was collected from Pasighat, Pangin and Mebo in the month of April, 1995, March, 1996. It was growing in forests soil under trees.

Habit

Solitary (Figure 6.13); Atm. Temp–21°C, R.H.–90 per cent, Soil Temp–23°C, Soil moisture-80 per cent, Soil pH–6.2.

Amanita vaginata (Fries) Vittadini

Pileus; 4.5-9.5 cm in diameter; first ovate then expanded, orange brown or grey or dark grey, texture; soft, fleshy context 10-12 µm thick, surface usually smooth stipe whitish grey, central, cylindrical, long and slender 9.0-15.0 cm long, 0.6-1.0 cm thick, ring absent, volva; tall and sack like, whitish grey, basidia clavate 30.50 × 10–13 µm tetrasporic, basidiospores white, globose, non amyloid, 9-11 µm in diameter, spore print; white.

Habitat

It was collected from Pangin in the month of March, 1995 and from Mebo in April, 1996, solitary, common in woodlands and forests.

Habit

Solitary (Figure 6.14) Atm. Temp–22°C, R. H.-91 per cent, Soil temperature–24°C, Soil moisture 83 per cent Soil pH–6.1.

Volvariella bombycina (Fries) Singer

Pileus; 12.0-20.0 cm in diameter ovoid–convex when young, later expanded, whitish, creamy whitish, surface; fibrillose, covered with silky hairs; soft, fleshy, context; 10-15 µm thick, stipe; firm and solid, often curved unto 20 cm long gills; crowded and free, white at first then pink, volva; very large, persistant, and some what viscid, flesh; white, soft, basidia; 4 spored, cystidia–abundant, basidiospores; pink, 7–8 × 5–6 µm spore pink.

Habitat

It was collected from Hydel Pasighat in the month of August, 1996 and also from Balek, in Sept. 96; usually growing on decomposing stumps in and around forests

Habit

Solitary (Figure 6.2) Atm. Temp-28°C, R.H.–89 per cent.

Pleurotus sajor-caju (Fr.) Singer

Pileus; 6.0–15.0 cm in diameter: oyster shaped to deeply infundibuliform, coralloid appearance, lobed and folded, colour grey or dull brown or greyish ochre, texture; soft and fleshy; surface smooth, margin; irregular and incurved, stipe eccentric, fleshy in fresh, origin on drying 1.5–3.0 cm long, 0.1–0.2 cm thick, gills decurrent, white in fresh, yellow in dry, basidia clavate 35–55 × 5–12 µm, basidiospores hyaline, cylindrical 7.0–7.5 × 3–3.5 µm, non–amyloid.

Habitat

It was collected from Basar and Ledum in the month of August–September, 1995 and July, 1996; also from Roing in July, 1996. Lignicolous, on decaying wood.

Habit

Solitary or in groups, Atm. Temp–29° C, R.H.–78 per cent to 81 per cent.

Pleurotus conrnucopiae (Paulet) Rolland.

Pileus; 80–150 µm in diameter, convex when young, later expanded and depressed, reddish brown or tinged pink or lilac or grey fading to greyish brown in age, texture; tender-fragile, surface; smooth, dry, margin smooth-often splitting, stipe; usually 2.5–5.2 cm long, eccentric, white or yellowish may be brownish, gills; crowded at cap margin, decurrent showing net like pattern, flesh; white, tender thick and soft, basidiospores; oblong, white 8–12 × 3.5–4.5 µm, spore mass pale lilac, spore print; pale-lilac.

Habitat

It was collected from Sille, Mebo and Pasighat in the month of June, 1995, July, 1996, August, 1996. It was found on old trunks of dead tree or on dead portion of living tree trunks.

Habit

Clustered–Gregarious (Figure 6.4). Atm. Temp–28°C to 29°C, R.H.–78 per cent to 80 per cent.

Pleurotus squarrosulus (Mont) Sing.

Pileus; 2.0–7.0 cm wide, convex or parabolic when young, sub-infundibuliform with a centric depression when mature, colour; whitish to cream colored, fleshy and bracket like, surface; smooth, minutely scaly; gills; crowded, decurrent, stipe; more or less eccentric or central, 2.0–4.5 cm long flesh; white spore print; white, basidia; clavate, 12.6–17.5 × 3.0–4.5 µm, basidiospores; hyaline, oblong elliptical, 4.6–6.0 × 2.8–3.5 µm.

Habitat

It was collected from Hydel Pasighat in the month of July 1996, also from Balek, Pasighat doimukh and Itanagar in August 1996. It was fairly common growing on the decaying logs.

Habit

Clustered in over lapping tiers. Atm. Temp–29°C to 31°C. R. H.–78 per cent to 88 per cent.

Pleurotus ostreatus (Fries) Kummer:

Pileus; 5.0–12 cm, expanding to convex when young later expanded and depressed, colour, whitish or tinged pink or ochre brown variable, gills; crowded decurrent; texture; fragile, glabrous; surface; smooth, stipe; eccentric or lateral, short or virtually absent, flesh; white, soft, thick and tender, basidia; four-spored, clavate, basidiospores, white sausage-shaped, 10–11 × 3.5–4.2 µm, non-amyloid, spore print; lilac.

Habitat

It was collected from Mebo, Tato in the month of June, 1995 and July, 1996. It was growing on dead wood or decaying logs.

Habit

Clustered in over lapping tiers (Figure 6.1), gregarious. Atm. Temp–27°C–28°C, R. H.–75 per cent–89 per cent.

Pleurotus fossulatus (Cooke) Sacc.

Pileus; 3.0–14.0 cm in diameter, plano-convex when young later flabelliform; colour; whitish to creamish, surface; at first glabrous, rimose when mature, margin involute when young, revolute at maturity, stipe; more or less eccentric, often lateral, 4.0–7.0 cm long, white, flesh; white, brittle on drying, gills–crowded, decurrent, anastomosing towards the stipe, basidia; tetrasporic (4–5 spored), 6.0–11.0 µm broad, basidiospores; ellipsoid cylindrical non amyloid, hyaline, 8-12 × 4.5–6.0 µm.

Habitat

It was collected from Mebo, Tato in the month of July 1996 and June 1997. Fruit body was usually growing on dead wood or decaying logs.

Habit

Clustered in groups Atm. Temp. 28°C–29°C, R. H. 77 per cent–87 per cent.

Pluteus cervinus (Fries) Kummer

Pileus; 5.0–12.0 cm in diameter, in young stage companulate, then convex and finally expanded, colour; white, yellowish brown or sooty; texture; soft, surface; radially fibrillose, scaly, stipe fairly long, solid and equal or sometimes swollen at the base 6.0–15.0 cm long, whitish, looks grey due to dark fibrils, gills; crowded and free, white at first then pink, finally brownish, basidia, 4-5 spored, 11.6–16.0 × 3.4–4.2 µm, basidiospores rosy pink, broadly elliptical and smooth, 7–8 × 5–6 µm.

Habitat

It was collected from Pasighat and Roing in the month of June, 1997, July, 1997. Fruit body was usually growing on rotten stumps and fallen trunks or on saw dust.

Habit

Solitary (Figure 6.5) or sometimes gregarious, Atm. Temp–26°C–27°C R.H.–89 per cent–90 per cent.

Crepidotus mollis (Fries) Kummer

Pileus; 1.0–7.0 cm in diameter, shell or kidney shaped, colour, pale brownish or yellowish, texture; soft, glabrous, surface, with epicuts composed of hyaline repent parallel hyphae, stipe; rudimentary and lateral, gills; crowded and decurrent to central point, whitish, then watery, cinnamon often spotted basidia; 4–spored. Cylindrical 21.00–25.0 × 5.0–7.0 µm, spore print; brown basidiospores; smooth and elliptical 6.5–8.0 × 4.5–6.0 µm.

Habitat

It was collected from Pasighat, Mebo and Roing in the month of June 1996 and July 1997. It was often growing tiered on stumps and decaying woods.

Habit

It groups, clustered (Figure 6.6). Atm. Temp–27°C-28°C. R. H.-89 per cent–91 per cent.

Flammulina velutipes (Fries) Karsten.

Pileus; 2.0–6.0 cm in diameter, convex to flattened, fleshy at disc but thin at margin, colour; tawny yellow, darker at disc, smooth and viscid, surface; glabrous, viscid, margin; inrolled, stipe; tough, cartilaginous, central, 3.0–9.0 cm long, 0.3–1.0 cm thick, pale, yellow covered with reddish brown hairs without volva and annulus, gills, adnate or adnexed, sub distant and very unequal, yellowish white, broad round near the stipe, basidia; tetrasporic, basidiospores, white, elliptic 6.5–10 × 3–4 µm and spore print white.

Habitat

It was collected from Mariyang and Tuting in the month of March, 1995, 1996. Growing on living or dead wood.

Habit

Caespitose, growing in clumps (Figure 6.7), Atm. Temp. 11°C–12°C; R.H.–87 per cent–91 per cent.

Coprinus micaceous (Fries)

Fruit body (Pileus); size 2.5–4.0 cm, ovate then campanulate and finally expanded, colour; ochre brown, but greyish at margin, texture, tender and delicate, surface; scaly, glistering scales on the surface, stipe thin hollow and dirty white, 6.0–9.0 long 0.4–0.7 cm thick, gills; adnate, close and lanceolate, at first white finally dark brown, basidia; polymorphic having broad sterile cells, basidiospores; brown to black lemon shaped 7.0–9.5 × 5.0–7.0 × 4.5–6.0 µm.

Figure 6.1: *Pleurotus ostreatus* (Fries) Kummer

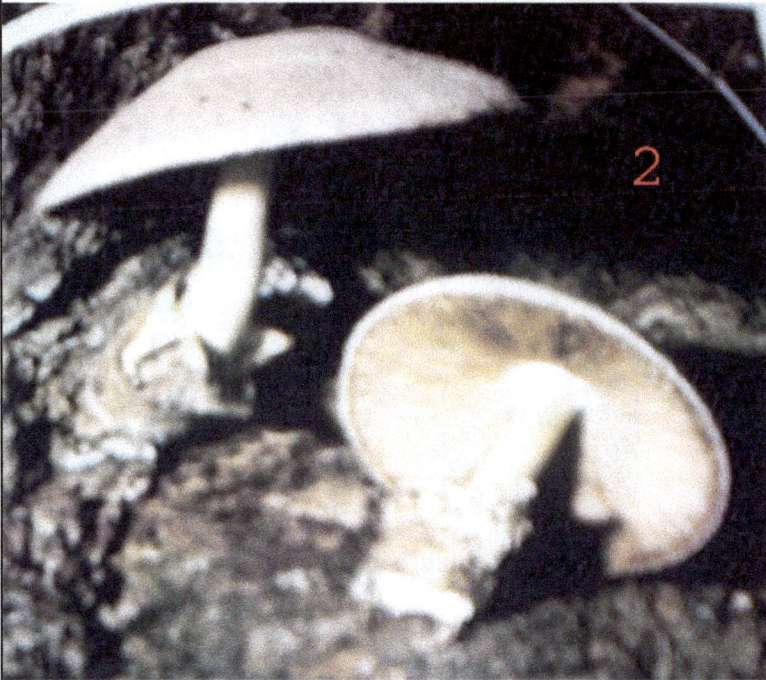

Figure 6.2: *Volvariella bombycina* (Fries) Singer

Figure 6.3: *Agaricus arvensis* Schaeff ex. Seer.

Figure 6.4:
Pleurotus connucopiae
(Paulet expers)
Rolland

Figure 6.5:
Pluteus cervinus
(Fries) Kummer

Figure 6.6: *Crepidotus mollis* (Fries) Kummer

Figure 6.7: *Flammulina velutipes* (Fries) Kasrsten

Figure 6.8: *Coprinus micaceous* (Fries)

Figure 6.9: *Armillaria mellea* (Fries) Kummer

Figure 6.10: *Tricholoma caligatum*
(Viviani) Ricken

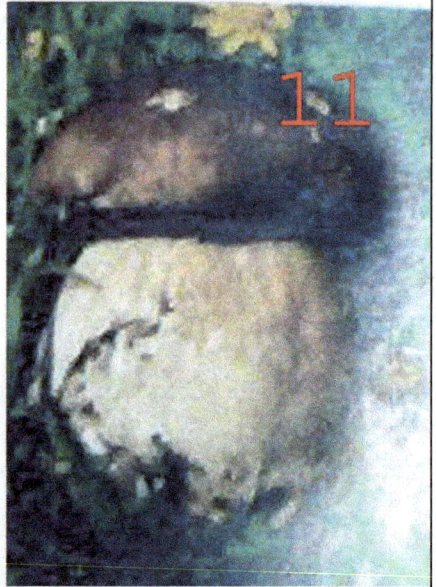

Figure 6.11: *Boletus edulis*
Bull-ex Fr.

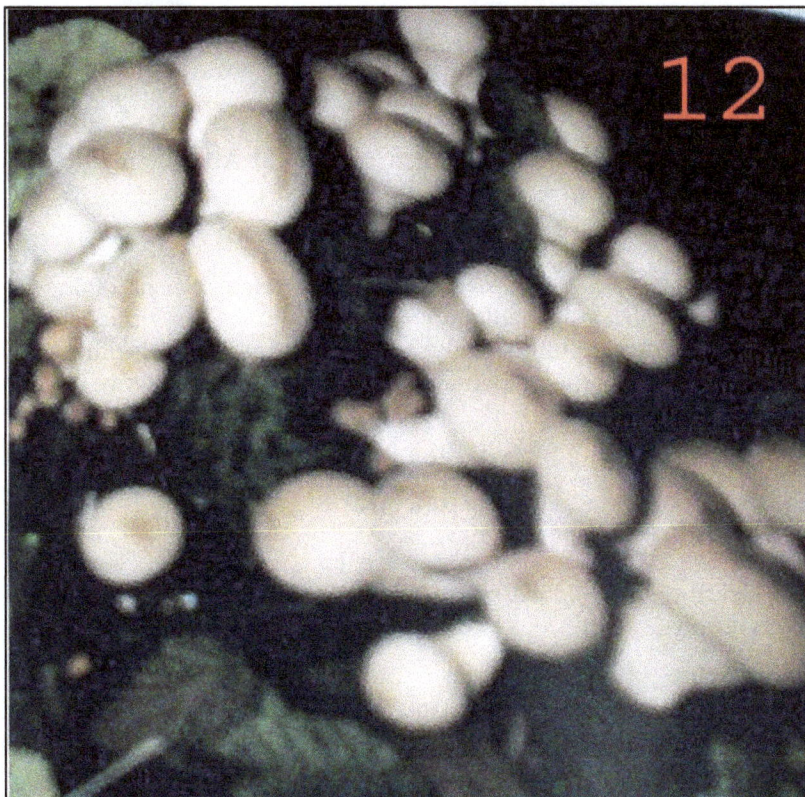

Figure 6.12: *Lycoperdom pyriforme* Persoon.

Figure 6.13: *Amanita caesarea* (Fries) Quelet

Figure 6.14: *Amanita vaginata* (Fries) *Vittadini*

Habitat

It was collected from Pasighat and Mebo in the month of April 1996 and July, 1996. Growing on old stumps and rotting timber.

Habit

In groups, gregarious in dense clumps (Figure 6.8). Atm. Temp–31°C, R.H.–86 per cent–90 per cent.

Armillaria mellea (Fries) Kummer

Pileus; 3.5–14.0 cm in diameter, convex then expanded, often becoming centrally depressed, margin striate, variable in colour olive tawny, grey pinkish or honey coloured reddish brown, texture; spongy soft, surface; scaly covered with dark colored scales; stipe; central uniform, rigid more or less grooved whitish above ring, reddish brown with age, ring whitish, flecked at margin with yellow or reddish brown, gills; sub-distant, adnate or slightly decurrent, whitish then pale brownish yellow, basidia; tetrasporic clavate basidiospores; whitish, elliptical 8–9 × 5–7 μm.

Habitat

It was collected from Ziro, in the month of January, 1996 and February, 1997. It was commonly found around the base of old trees on the ground in the forests forming thick rhizomorphs.

Habit

Gregarious, polymorphic, tiered around the base of old trees (Figure 6.9). Atm. Temp. 22°C–24°C, R.H.–85 per cent–89 per cent, Soil Temp–26°C, Soil moisture–74 per cent–84 per cent Soil pH–5.9.

Tricholoma georgii (Fries) Quelet.

Pileus; 5.0–8.0 cm in diameter, convex to expanded flat, slightly umbonate, colour, whitish, or cream coloured or light buff, surface smooth without any scale, hyphal epicutis, present, stipe; cylindrical solid later becoming hollow centric, fleshy, curved whitish, 3.5–8.0 × 0.8–1.0 cm, gills; sinuate or adnexed, crowded, white; pen-knife shaped, trama thick walled, regular, basidia; clavate, 4-spored, basdiospores; hyaline elliptical, 5–6 × 2–3 μm smooth, non amyloid.

Habitat

It was collected from Ziro in January,1995. It was humicolous found on lawns and forest floors.

Habit

In groups, gregarious, Atm. Temp.–22°C–25°C, R.H.–83 per cent–91 per cent.

Tricholoma caligatum (Viviani) Ricken.

Pileus; 8.0–18.0 cm in diameter, convex, later expanded, finally plane, colour; yellowish–ochraccous or reddish-brown, surface; silky aquamulose; stipe; tall equal, tapering downwards, white; gills; crowded sinuate, narrow white, eroding and becoming free, basidia; clavate, four spored, basidiospores; white, elliptical and smooth, 6–7. 5 × 4.5–5.5 μm in amyloid, cystidia absent, flesh; thick firm and white.

Habitat

It was collected from forests near Ziro in the month of January, 1995. Gregarious under coniferous trees in forests.

Habit

Gregarious (Figure 6.10), in groups, Atm. Temp–24°C–26°C R.H.–84 per cent–89 per cent.

Boletus edulis **Bull-ex. Fr.**

Pileus; 9.0–18 cm. broad, hemispherical then convex when young, broadly convex with age, colour; variable, yellowish brown, context; firm, surface; dry; viscid when wet. Glabrous, smooth, uneven, wrinkled to some what shallowly pitted, stipe; central 40–90 μm long. 10–20 μm across, often swollen in the middle or below, irregularly shaped, reticulate in the upper half, reticulatus white, flesh firm, white tubes, adnexed, long and thin, pale whitish when young, yellowish white to olive yellow in age, easily detached from cap, pores; minute and roundish, stiffed when young, yellowish white to olive yellow in age, basidia, clavate, 4-spored, basidiospores; 12–16 × 4–5.5 μm, bronze coloured, fusiform, yellowish to olive yellow in KOH.

Habitat

It was collected from forests near Ziro, Pangin and Boleng in the month of March, 1995, and April, 1997. It was common on the ground in open forest or in gardens.

Habit

Scattered (Figure 6.11) to gregarious, sometimes solitary, Atm. Temp–24°C–25°C, R.H.–88 per cent to 89 per cent, soil Temp–25°–26°C Soil moisture 76 per cent–79 per cent, Soil pH–5.8–5.9.

Boletus scaber **(Fries)**

Pileus; 60–100 μm, convex to parabolic, surface dry, viscid when wet, glabrous and often with depressions in age, light grayish brown in the centre, fading to yellowish brown towards the periphery, margin; even and not projecting beyond the tubes, Context; 10–12 μm, white, tubes; 6–15mm deep, deeply depressed around the stipe; pores minute, 2-3 μm, roundish, whitish first later brownish, stipe; 60–120 μm long, 15–25 μm broad at base, gradually tapering upwards, yellowish white black scabers, flesh; whitish yellow to pale yellow, basidia; clavate, four spored, 27–40 × 12–15 μm, Basidiospores, 15–19 × 5 × 7 μm, elliptical to subfusiform brownish, dextrinoid cheilocystidia 30–55 × 8–12 μm, pleurocystidia 55–78 × 11–75 μm.

Habitat

It was collected from Mariyang, Tuting in the month of July, 1995. It was found under trees in forests.

Habit

Solitary to scattered, Atm. Temp.–24°C–25°C, R. H.–89 per cent–91 per cent.

Suillus sibiricus (Singer) Singer.

Pileus; 2.5 to 10.0 cm in diameter, convex to plano convex, pale yellowsih when young to olive yellowish in age, context; 4–6 µm deep, firm, surface; viscid to glutinous, spotted, overall covered by brownish appressed scales, stipe; 3.0-8.5 cm long 7-15 µm wide, solid pale yellow, then yellow brown, glandular dotted overall, occasionally with a ring glandulate darken in mature specimens, tubes; adnate, turning decurrent 5-10 µm deep pale yellowish to yellowish, pores; radially arranged, angular, 1-2 µm broad, pale yellow to yellowish brown, basidia; 4-spored, clavate, basidiospores; light brown, 8-11 × 3.5-4.0 µm, narrowly elliptic, inamyloid.

Habitat

It was collected from Ziro in the month of January, 1996 and February, 1997. It was found on the soil under coniferous forest.

Habit

Gregarious, Atm. Temp 21°C–22°C. Soil Temp–24°C–25°C. R.H. 81 per cent–89 per cent, Soil moisture; 78 per cent–79 per cent, Soil pH–5-9.

Leccinum oxydabile (Singer) Singer.

Pileus; 20-90 µm broad, globose to parabolic when young, expanding to convex to broadly convex when mature, reddish brown in young, then light brownish, surface dry, viscid when wet; smooth to finely tomentose, margin regular smooth, context; 8–12 µm thick firm, tubes 15-20 µm deep, deeply depressed around the stipe, pores; roundish, yellowish brown, stipe central, 60–125 µm long gradually tapering upwards, basidia clavate, four spored, basidiospores; brownish, elliptical, 16–19 × 5–6 µm elongate, inamyloid, occasionally dextrinoid.

Habitat

It was collected from Maryang, Tuting in the month of January, 1996, Ectomycorrhiza *i.e.* with trees in forests.

Habit

In groups of two to three, Atm. Temp–23°–24°C, R. H. 88 per cent–91 per cent.

Jelli Fungi

Hirneola auricula–Judae (Bull. St. Amans) Berkeley.

Fruit body; 2.5 cm when young, at maturity usually 7.0–9.0 cm, ear shaped or irregular cup shaped, translucent, jelly like or gelatinous tough when dry, surface velvety with greyish hairs, yellow brown to reddish brown when moist, five distinct hyphal zones in transverse section, hymenium smooth, veined and reddish 145–155 µm wide, basidia cylindrical, transversely septate, 50–60 × 5–6 µm., basidiospores; white oblong, curved and narrow at the base, 14–19 × 5–9 µm.

Habitat

It was collected from Pasighat, Mebo, Ruksin and Bilat in the month of March, 1995; April 1995; August, 1996 and July, 1997, on wood (Lignicolous), or on dead wood or logs, attached to the branches of trees.

Habit

Solitary or gregarious in dense tufts, Atm. Temp. 26°C–28°C, R.H.–86 per cent–88 per cent.

Hirneola polytricha Mont.

Fruit bodies; 2.0–3.0 when young at maturity 8.0–10.0 cm, cup shaped or ear shaped, red brown and rubbery gelatinous when fresh, grey to tan brittle, cartilaginous on drying, surface velvety, eight distinct hyphal zones in transverse section, hymenium; 80–105 µm. wide smooth and papillate, basidia cylindrical transversely septate; basidiospores oblong, curved 14.5–16.0 × 4.5–9.0 µm.

Habitat

It was collected from Mebo and Tezu in the month of July, 1996 and 1997 respectively. It was found to be lignicolous on dead branches of Ficus species.

Habit

Solitary, Atm. Temp–27°C–31°C, R. H.–84 per cent–91 per cent.

Auriculuria fuscosuccinea (Mont) Farl.

Fruit bodies 8–12 cm wide, cup shaped, colour rosy to vinaceous when fresh, translucent when dry, sessile, sometimes substipitate, texture; cartilaginous, surface; pileate, upper surface grey and hairy, hyaline hairs with round tips; hymenium; smooth 70–80 µm wide; basidia cylindrical 45–60 × 4–6 µm wide; basidiospores; allantoid 14–16 × 6–7 µm in size.

Habitat

It was collected from Roing in the month of July, 1996 and March, 1997. It was tenaceously attached to the dead branches of trees.

Habit

Solitary or gregarious, Atm. Temp–28°C–31°C, R.H.–89 per cent–91 per cent.

Auricularia delicata (F.r.) Henn.

Fruit bodies; 5.0–8.9 cm wide, ear shaped, mostly sessile, colour vinaceous grey, margin slightly curved, texture rubbery, gelatinous; tough; surface with hyaline hairs; six distinct hyphal zones in transverse section; hymenium merculoid to strongly poroid reticulate, 80–95 µm wide; basidium cylindrical transversely septate. Basidiospores allantoid 12–14 × 5–7 µm in size.

Habitat

It was collected from Roing in the month of July, 1996 and Pasighat in the month of August, 1997. Lignicolous, growing on wood.

Habit

Solitary or gregarious, Atm. Temp–28°C to 31°C, R.H.-86 per cent–91 per cent.

Lycoperdon pyriforme Persoon.

Fruit bodies; 3–4 cm high, 2.0–4.5 cm broad, pear shaped, colour, whitish, brownish or yellowish; texture soft; surface, covered with minute reddish–brown pointed warts and scales, these erode with maturity leaving a smooth yellowish surface; peridium cracking during wet weather, capillitium–initially greenish yellow but later on dull, olivaceous; columella present; basidiospores olive or greenish yellow, smooth and spherical, 3.3–4.5 μm in diameter.

Habitat

It was collected from Mariyang and Along in the month of April 1995 and March-April 1996. It was growing on decaying wood or on the forest ground but always attached to wood.

Habit

Solitary or crowded together (Figure 6.12), Atm. Temp.–20°C–22°C; R.H.–91 per cent–92 per cent.

Lycoperdon perlatum (Persoon)

Fruit bodies; 3–6 cm across, 4–8 cm high, pear or top shaped, colour white or dirty or whitish becoming yellowish brown; texture soft; peridium spherical with a stem like base; surface covered with short fragile spines; spines usually falling off at maturity, capillitium, at first greenish yellow with on olive tinge, later pale brown; columella present; basidiospores olive to pale brown, smooth and spherical, 3.5–4.5 μm in diameter, released through an apical pore.

Habitat

It was collected from Pangin and Mariyang in the month of March 1995. Scattered on the ground in forests or in open places.

Habit

Usually gregarious, Atm. Temp.–19°C–21°C, Soil Temp.–22°C–24°C, R. H.–89 per cent–91 per cent, Soil moisture–76 per cent, Soil pH–6.2.

Langermannia gigantea (Person) Rost Kovius.

Fruit bodies; 20–45 cm in diameter, round or globose or slightly flattened on the top colour whitish when young, then yellowish to olive brown; surface; smooth or may by covered with fine scales, cracking flesh or gleba firm and white at first, then yellowish, finally olive brown and pulverunt; stipe absent or only present as a small come of tissue; capillitium yellow and finally dull olivaccous; spore mass whitish cream coloured, and finally olive, brown; basidiospores brownish, spherical, 4-5 μm. billions of spores released through cracks which start at the top.

Habitat

It was collected from Balek, Village and Rani village, near Pasighat, in the month of April, 1996. It was common in pastures, gardens.

Habit

Solitary, rarely gregarious Atm. Temp–25°C–26°C R.H.–85 per cent, Soil Temp–27°C–28°C, Soil moisture–75 per cent–78 per cent, Soil pH–6.1.

Calvatia Cyathiformis (Bose) Morgan

Fruit bodies; 8–16 cm in diameter, rounded or pear shaped with stalk, colour white when young, pinkish brown when mature; peridium surface cracking into irregular and brittle fragments, flesh or gleba white when young purplish at maturity. Capillitium purple brown; basidiospores globose to subglobose, purple brown, echimulate to echinate, rough, 5.0–6.5 μm, spore mass dark purple with age.

Habitat

It was collected from Sille village and Mebo in the Month of April 1996. It was commonly found in grassland or cultivated lands.

Habit

Gregarious, in groups, Atm. Temp–28°C–29°C, R. H.–75 per cent–85 per cent, soil Temp–30°C–31°C, Soil moisture–74 per cent–79 per cent, Soil pH–6.1.

Podaxis pistillaris (L. ex. Fr.) Morse.

Fruit bodies; 20–25 cm across, pyriform or rounded cylindrical, colour dirty white or fawn; texture hard and woody, surface; breaking into scales; endoperidium, firm and rigid, dehiscing by splitting, sometimes exposing the powdery gleba; copius with reddish brown spore mass but dispersed with in capillitium at maturity; capillitium aseptate, unbranched and coiled; basidiospores reddish brown, ellipsoid to subglobose, thick walled covered by a gelatinous sheath with a pre-eminent germ pore, 7–21 × 5–15 μm.

Habitat

It was collected from 21st. mile near Pasighat in the month of July, 1997. It was growing on sandy soil or mud.

Habit

Suberranean, above ground when ripe, Atm. Temp–29°C–30°C, R.H. 76 per cent–78 per cent, Soil moisture–74 per cent–76 per cent, Soil pH–6.6 6.9.

Earth balls

Scleroderma aurantium Persoon.

Fruit bodies; 4.0–8.0cm in diameter, roughly oval but often uneven; colour; olive yellow to brownish, peridium surface; scaly often in a net like manner, pendulum wall thick olive yellow to brownish, gleba; greyish to purplish black, glebal cavity traversed by white tramal plates, spores are embedded within amatrix; capillitium, absent, rudimentary; basidiospores in mass are black with a purple tinge, reticulated and globose, blackish brown, 12–16 μm, spores escape through cracks in the peridium.

Habitat

It was collected from forests, near Mebo and Gelling in the month of April, 1997. It was found in forests under trees.

Habit

Solitary rarely gregarious, Atm. Temp–20°C–21°C, R.H.–79 per cent–86 per cent.

Scleroderma verrucusum (Bull) Persoon.

Fruit bodies; 4–9 cm in diameter, flatish ovoid or subglobose, colour, greyish to light brown; peridium finely warted, cracks irregularly to liberate spores, mycelial cords showing more or less like a stipe; gleba umber coloured, glebal cavity traversed by tramal plates; spores are embedded in a matrix when young, capillitium rudimentary or absent; basidiospores dark brown with prickles, 10-13 μm.

Habitat

It was collected from Mebo, Mariyang and Jenging, in April 1997. It was found on waste ground, in forest soil or in sandy soil.

Habit

In group, gregarious, Atm. Temp–22°C–24°C; R. H–80 per cent–88 per cent.

Polypores

Polyporus brumalis (Pers) Per, Fr.

Pileus; 2–5 cm in diameter, convex at first but finally depressed; colour, greyish to light brown, older specimen's develop darker concentric zones, surface, hairy minutely scaly without zonations, stipe white or dark brown, tending to be darken at the base, central, slightly eccentric and very tough; pores, circular or slightly angular, tubes; short and decurrent, up to 0.25 cm long, basidia club shaped, 4–5 μm wide, basidiospores hyaline, oblong 5–8 × 4.5–2.5 μm.

Habitat

It was collected from Pasighat in the month of March, 1995. It was attached to wood of frondose tree stumps and roots and also on half burried log of *Betula, Shorea* etc.

Habit

Grows singly or a few together, Atm. Temp.–26°C–28°C, R. H.–86 per cent–89 per cent.

Laetiporus sulphureus (Bull, ex. Fr.) Murrill.

Pileus, 30 to 40 cm fan or tongue shaped with the margin undulate and lobed; colour–orange yellow or sulphur yellow margin, later fading becoming dull and dirty, surface; smooth, sometimes with concentric furrows, stipe, very short or absent as such, the many overlapping brackets arising from a common base on the host, pores; elliptic or circular, tubes short and sulphur yellow; basidia; club-shaped, 4-spored, basidisopores hyaline, elliptical 5–7 × 4–5 μm, thin walled, whitish.

Habitat

It was collected from Ziro and Mariyang in the month of January, 1995. It was attached to dead wood of coniferous trees.

Habit

Usually gregarious, Atm. Temp.–22°C–24°C, R. H.–87 per cent–91 per cent.

Cup fungi

Sleuria aurantia (Pers. ex fries) Fuckel.

Fruits bodies; short, 1.5–4.0 cm in diameter, at first cup shaped but flattened with age, sessile or laterally stipitate, sometimes split along the margin, flesh; thin and brittle turning green by iodine, hymenium; bright orange; asci 8-spored, not turning green by iodine 132–171 × 14.0 µm. ascospores, elliptical, surface reticulate, presence of oil drops 13.2 × 10.5 µm paraphyses clavate, projecting beyond asci.

Habitat

It was collected from Pangin, Tuting in the month of March 1995. It was growing on ground or on bare sandy soil.

Habit

In groups, gregarious, Atm. Temp–22°C–23°C, R.H.–91 per cent–92 per cent.

Ethnomycological Observations

Arunachal Pradesh is mainly a tribal state. The total number of tribes in the state is around hundred and ten. Besides local tribal population, Nepalis are also inhabiting the state for the last hundred years or more. The important tribes of Arunachal Pradesh are Adi, Nishi, Monpa, Apatani, Sherdukpens, Miji, Bangni, Aka, Tagin, Hill miri, Momba, Khamba, Mishmi Nocte, Khampti, Singpho, Wangcho, Yobin and Tangsa etc. Monpas and Sherdukpens are found in Kameng district just bordering Bhutan. Apatanis, Nishis and Tagins are inhabiting in the Subansiri area of the state. The Apatanis are inhabiting in the Ziro valley at an altitude of approximately 1524 meters. Siang is the home of several tribal groups but predominantly of the Adis, until a few years back called Abor. The tribes living along the northern borders are the Tagins, Membas and Khambas.

The Adis of Siang fall into two broad divisions, the first comprises Padams, Minyongs, Pasis, Panggis, Shimongs, Boris, Ashings and Tangmas and the second includes Gallongs, Ramos, Bokars and Pailibos. The Lohit district is the home of the Mishmis, the Khamptis and the Singphos. Tirap is the home of the Wancho, the Nocte, and the Tangsa. The main concentration of the Tangsas is in the valleys of the Tirap river and the Namchik extending from Patkoi range in the south to the border of the state of Assam in the north.

The tribals are well acquainted with the utility of mushrooms and have given different local names to them. The Pangi clan of Adis calls small edible mushroom such as *Lycoperdon* is called as Tapar. The large edible mushrooms which are found generally growing on woods such as *Pleurotus, Pluteus, Auricularia* and some Polypores are called as Tadar. The mushrooms which grow in soil are called as Indeng. The mushrooms which are found on dung are locally known as Intimomi. The Adis tribe near Pasighat calls edible mushrooms as Tapar. Adi Padam of Maryang calls a mushroom Lolum when it grows on wood during rainy season. Minyong clan of Adis also calls the mushrooms as Indeng or Tatar to the mushroom growing on wood log.

The Gallongs of West Siang call mushrooms as Tuying. In few places it is called as Atar, Hubsi is the name given to smaller edible mushrooms by Gallongs, which grow in soil. The Adis at Belong called every mushroom as Tatar. The Nishi tribe of subansiri and Apatanis of Subansin district calls mushrooms as Tuying. In some locality it is called as Tapin. The Apatani people inhabiting near Ziro calls it Tain. Tain was generally large edible mushrooms while Tuying were smaller forms.

The Hill Miris calls mushrooms as Hatar or Atar. The Tangsas of Tirap call pusoublog to different mushrooms. The sherdukpens of West Kameng call mushrooms as Mi., the clan of Tangsa Hachumg as Kangkuirang. The Tagins of Upper Subansiri as Tin and the subclan of Tangsa, Luzgchang as lok. The non-edible toadstools are called as Yyitin by Tagins. Sang is applied to mushroom growing on tree while thin is applied to the mushrooms growing on soil.

The Khampti tribe of Lohit area calls edible mushrooms as Nohew. The Nepali community living in Arunachal Pradesh calls mushrooms as cheo. The mushrooms growing on wood or log or tree trunk or branch are called as Kath cheo. The mushrooms which are growing in soil are called as Mati cheo. The mushrooms which are found in bushes or in grassland community in open sunny place are called as Jhar Cheo. The Bengali community of Arunachal Pradesh calls mushrooms as Banger Chhata or Dudh Chhata.

Tribals use mushrooms as their diet in four ways. Some very delicious mushrooms such as *Pleurotus* species or *Agaricus* species; they mix it with vegetables and local green leaves. This mixed vegetable preparation of mushrooms is very delicious dish for them and if taken for some time it improves the health. According to their opinion 250gms. to 500 gms. of fresh mushroom is sufficient for one meal of small family. Others consume mushroom in boiled form. In this preparation the fresh mushrooms are first washed and then it is boiled in water on mild heat for 5–15 minutes over heating of mushroom spoils its taste. The boiled mushrooms are then mixed with some salt and spices and taken as nutritious food. Fried mushrooms are also used by local population. In this preparation mushrooms are cut into pieces and then fried in saucepan with some other ingredients. Ingredients may be salt, pepper, fine white bread curbs and other spices. Some mushrooms are eaten raw by the local people. Stipe of the mushroom is more tasty compared to its pileus if taken raw according to their information.

Some species of mushrooms are not consumed because of their bitter taste. Some of the villages consume these bitter mushrooms by applying some old age method. In this method, the mushrooms are covered in a leaf of some local plants and they are placed in a dig for some time after covering it with soil. After two to four days, they take it out from the soil and boil in water for half an hour. The water is then removed from the pot and boiled mushrooms were consumed with salt and spices.

Prospect of Mushrooms Cultivation in Arunachal Pradesh

Chandra *et al.* (1996) studied the possibility of year round cultivation of oyster mushroom (*Pleurotus* spp.) under natural environment for North-East region. Six different types of oyster mushroom are tested at Barapani, Meghalaya. Three strains of *P. flabellatus*, two strains of *P. ostreatus* and one strain of *P. sajor caju* are grown and tested in different months. They concluded that *P. flabellatus* strains are better adapted for wide range of temperature, followed by *P. sajor caju* which performed well except in cold season. *P. ostreatus* is suitable for cultivation during cooler months of the year. Thus, oyster mushroom, which grows on various agricultural wastes and being adopted world wide, can be taken up for commercial cultivation in Arunachal Pradesh also (Tables 6.2 and 6.3).

Table 6.2: Production of Oyster Mushroom (*Pleurotus* sp.) in some Countries During 1994 (X1000 MT Fresh)

Country	Production
China	654.00
Japan	20.8
S. Korea	57.9
Thailand	15.0
Taiwan	4.6
Indonesia	1.0
U.S.A	0.9
Hungary	4.5
Others	38.7
Total	797.40

Source: Jandaik (1997), Chang (1996).

Huge Quantities of lignocellulosic wastes are generated annually through the activities of agriculture, forest and agro-based industries in Arunachal Pradesh. *P. sajor caju* and other species of *Pleurotus* can be grown well on these lignocellulosic waste materials (Rajrathnam *et al.* 1997). *Pleurotus* cultivation does not need any high tech requirement. This is one of the most popular mushroom grown worlds wide and its produciton has increased drastically (Chang, 1996; Jandaik; 1997). *P. sajor caju* is grown on decaying logs at the foot hills of the Himalayas. The higher humidity and low temperature of this area is very suitable for cultivation of mushroom. Its higher yield (179g/log) has been recorded on *Mangifera indica* followed by *Artocarpus lakoocha* and the lower yield *Casurina equisitafolia*. *Artocarps* species is very common in Arunachal Pradesh and hence and can be suitable substrate for the cultivation of mushroom.

The ecological and agroclimatic conditions of Arunachal Pradesh seem to be also suitable for the cultivation of button mushroom (*A. bisporus)* and *Flammulina velutipes*. The low temperature and high relative humidity of high altitude temperate areas are much suitable for their cultivation. Foot hills enjoying sub-tropical, tropical climate is also most suitable for cultivation of mushroom during winter months.

Table 6.3: Yield of Mushroom Species on Various Lignocellulosic Substrates

Substrate	Species	Yield kg per 10 kg Prepared Moistened
Banana leaves	P. quebeca	1.20
Bean Straw	P. pulmonarius	2.60
Bean Straw	P. cornucopiae	2.25
Corn cobs	P. hybridius	2.70
Corn cobs	P. citrinopileatus	1.90
Cotton Straw	C. velutipes	1.54
Flax rind	P. hybridius	2.42
Gram chaff	Stropharia regosoannulata	3.40
Composted horsemanure	A. bisporus	2.80
–do–	Lepiota naucina	1.45
Oil palm pericarp	P. hybridius	1.94
Pea nut hulls	P. sajorcaju	0.97
Rice Straw	P. citrinopileatus	2.40
Saw dust	C. velutipes	2.62
(enriched with bran)	Lentinus edodes	1.91
Shreeded paper	Pholiota mutabilis	1.64
Sorghum paper	P. hyridius	0.38
	P. sajor caju	1.60
Sugarcane bagasse	P. hybridius	2.30
Tree bark composted	P. sajor caju	0.97

Source: Poppe and Hofte (1995), Rajrathnam *et al.* (1997).

References

Ainsworth, G.C., 1971. *Ainsworth and Bisby's Dictionary of Fungi*, 6th Edn. C.M.I. Kew, England.

Chakravarty, D.K. and Sarkar, B.B., 1982. *Tricholoma lobayense*: A new edible mushroom from India. *Curr. Sci.*, 51(10): 531–532.

Chandra, S., Singh, A.K. and Kumar, S., 1996. Studies on the effect of temperature on mycelial growth of oyster mushroom. *Annual Report* (1995–1996), ICAR Research complex for NEH Region, pp. 67–69.

Chang, R.Y., 1996. Potential application of *Ganoderma* polysaccharides in the immune surveillance and chemo prevention of cancer. In: *Mushroom Biology and Mushroom Products*, (Ed.) D.J. Royseed. Penn. State Univ., Pennsylvannia, pp. 153–159.

Henderson, D.M., Orton, P.D. and Watling, R., 1968. *British Fungus Flora, Agarics and Boletii: Introduction.* HMSO Edinburgh.

Jandaik, C.L., 1997. History and development of *Pleurotus* cultivation in the world and future prospects. In: *Adv. in Mush. Biol. and Production,* MSI, NRCM, Solan, p. 46–52.

Lakhanpal, T.N., 1995. Mushroom flora of North West Himalayas. In: *Adv. in Horticulture, Vol. 13: Mushroom,* (Eds.) K.L. Chadha and S.R. Sharma. Malhotra Pub. House, New Delhi, pp. 351–353.

Lakhanpal, T.N., 1996. Studies in cryptogamic Botany. In: *Mushrooms of India: Boletaceae.* APH Publ. Corporation, New Delhi.

Largent, D.L., 1977a. *How to Identify Mushroom to Genus–I: Macroscopic Features.* Mad River Press Inc., Eureka.

Largent, D.L., 1977b. *How to Identify Mushrooms to Genus–II: Microscopic Features.* Mad River Press Ind., Eureka.

Malloch, D., 1971. Collecting mushrooms for scientific study. *Greenhouse Garden Grass,* 10: 78–82.

Poppe, J.A. and Hoffe, M., 1995. *Mushroom Science,* 51(7): 181.

Purkayastha, R.P. and Chandra, A., 1985. *Manual of Indian Edible Mushrooms.* Today and Tommorrow's Printers and publishers, New Delhi.

Rajarathnam, S., Shashirekha, M.N., Bano, Zakia and Ghosh, P.K., 1997. Renewable lignocellulosic wastes the growth substrate for mushroom production nation strategies. In: *Adv. in Mush. Biol. and Production,* MSI, Solan.

Singer, R., 1975. *The Agaricales in Modern Taxonomy,* Third Fully Revised Edn. J. Cramer, pp. 915.

Soothhill, E. and Fairhurst, A., 1978. *The New Field Guide to Fungi.* Michael Joseph, London, 191 pp.

Chapter 7

Biotechnological Renaissance in Horticultural Crops

Alka Srivastava and S.S. Raghuvanshi

In vitro Culture and Plant Genetics Unit,
Department of Botany, University of Lucknow,
Lucknow – 226 007

ABSTRACT

Biotechnology has manifested itself in cultivation and improvement of a large number of plant species including horticultural crops which in true sense require a proper overhauling for the improvement of yield and quality and also the production of secondary metabolites. Various techniques of *in vitro* culture *viz.* micropropagation, somatic embryogenesis, somaclonal variation, transformation and *in vitro* manipulation for excessive production of secondary metabolites have been used in recent times for the improvement of horticultural crops. Most of the experiments conducted upto date are aimed at establishing the protocol for future use. The results are of academic interest and their commercial application is possible only when large scale variants or transgenics are obtained for judicious selection of the desirable elite types. The present paper reviews some of the important developments along these lines.

Review

In India, about 72 percent of the total cultivated area is under food crops. A large variety of fruits, vegetables, flowers, spices, medicinal and aromatic plants, root and tuber crops cover roughly only about 7 percent of the gross cropped area but contribute more than 18-20 percent gross value of the agricultural output. India makes a valuable location for investment in horticulture and food processing sector. But besides a few major achievements, the growth of Indian horticulture has not been upto the mark. Maximum extension in area under horticultural crop cultivation has been done and for increasing the output, now emphasis is being laid down on developing marginal poor waste lands which are not suitable for economic cultivation of field crops. Besides increase in cultivated area, the application of the recently developed biotechnological techniques in order to increase the yield as well as produce new, unique variability is a must. The lack of availability of suitable propagules all around the year has proved to be a major drawback with regard to the extension of area under horticultural crops. Research laboratories all over the world are concentrating on the improvement of these crops using both conventional as well as the modern breeding methods. It is well known that seed propagation leads to the development of a totally heterozygous population which makes it difficult to maintain a uniform clone of the desirable elite type. Tissue culture, however, provides convenient, short duration alternative technique as it helps in the multiplication of a desired elite plant using only a small portion from it. In various laboratories work on this aspect is underway. Shoot tips of *Musa paradisiaca* cv Basrai formed multiple shoots when cultured on MS medium supplemented with 7 or 10 mg BAP per litre. Single shoots were obtained by culturing the explants on MS medium containing 5 mg BAP pet litre. Rooting was induced on MS medium with 2 mg IBA per litre and the plantlets were successfully transferred to pots (Raut and Lokhande, 1989). Swamy *et al.* (1988) cultured excised embryos of cv Pusa purple, long of *Solanum melongena* on MS medium supplemented with 0.5 and 1 mg of auxin, kinetin and benzylamine per litre in various combinations. Callus, roots and more than 50 per cent multiple shoots were induced on MS medium containing 0.5 mg 2, 4–D and 1 mg kinetin per litre. The shoots produced roots when transferred to NAA supplemented medium. In some plant species, success is not obtained on using explants from mature trees and therefore seedlings are first-raised. Seeds from 8 *Lycopersicon esculentum* cultivate and hybrids were germinated *in vitro* on hormone free MS medium by El-Farash *et al.* (1993). Hypocotyl and cotyledonary explants from 6, 12, 18 and 24 days old seedlings were cultured on MS medium containing IAA and kinetin. The rate of callus formation was 100 per cent from cotyledonary and hypocotyl explants of all genotypes. However, differences between genotype, explant type and age of explant donor were observed for shoot regeneration rate and number of shoot explant. Cotyledonary explants produced the highest number of shoots. The protocol is recommended as a high yielding and rapid micropropagation system for tomato.

In vitro clonal propagation of *Capsicum* sp. has been done by Christopher and Rajam (1994). The effect of various concentrations of BA and kinetin on shoot proliferation from shoot tip explants was investigated in *C. praetermissum* and *C. annuum*. Large number of shoots were obtained on MS medium supplemented with BA and kinetin in both species after 4 weeks of culture.

Garlic shoot cultures (Seabrook, 1994) were established *in vitro* using a small piece of basal plate, an axillary bud and one or 2 subtending bulb scales on a medium containing MS salts and 0.1 mg NAA/litre and 2.0 mg BA Bulb formation of garlic shoot cultures was obtained on a medium containing a lower concentration of macro and microelements. The mean numbers of shoots produced differed for different cultivars due to the genotypic differences.

In vitro propagation of *Mentha* species by apex culture and microcuttings was studied by Geslot *et al.* (1989) and a method combining the two techniques was proposed. It involves cultures of spices, excision of microcuttings from shoots formed on the apices, culture of microcuttings on MS medium containing IBA and BA and subculture at 6 week intervals.

Studies by Zirari and Lionakis (1994) have shown that shoot explants taken from avocado trees cut back two years previously showed higher *in vitro* shoot proliferation rates than explants from trees cut back in the current year. High shoot proliferation were also obtained with explants from cuttings of adult cv Duke trees. Shoot tip explants produced more shoots than single node segments. Attempts to root microshoots were unsuccessful due to severe shoot necrosis after transfer to rooting medium.

Frey and Janick (1991) have reported that shoot regeneration in carnation was influenced by genotype, explant source and plant growth regulator balance. Plants were regenerated from petal, calyces, nodes, internodes and leaves but only petals, calyces and nodes were regenerative. Maximum proliferation was achieved with petals on MS medium supplemented with 0.05 µM thidiazuron and 0.5 µM NAA. Shoot initiation originated from cells near vascular regions and perhaps from epidermal cells in petals and via organogenic callus from other explants.

Shoot multiplication of some olive cultivars has been tried by Seyham and Ozzambak (1994). Young shoots were collected from mature trees and segments with 3-5 axillary buds were placed in modified Rugini olive medium containing BA at 0.5 mg/litre. After 35-40 days, explants showing good growth were transferred to a proliferation medium containing 0.5, 1.0 or 2.0 mg BA and 0.05 mg NAA/litre. The maximum number of healthy shoots was induced with BA at 1.0 mg/litre. The common problem of exudition of toxic phenolic compounds by explants during the first two-three days of culture could be overcome by collecting shoots during June to September when exudation was at its lowest, by selecting shoots with a diameter of 2-4 mm and surface sterilizing them and by soaking shoots in sterile distilled water before culturing them in dark.

Efficient *in vitro* regeneration of leek (*Allium ampeloprasum*) has been done by Silvertand *et al.* (1995) using flower stalk segments of 29 genotypically different plants and culturing them on MS medium with BA and NAA. The percentage of regenerated explants (PREX) ranged from 10-100 per cent and average number of regenerated shoots/explant (NSHO) from 0-40 for the 29 plants.

Shoots were regenerated from cotyledons of matured stored seeds of three peach rootstock cultivars by Pooler and Scorza (1995). Shoot regeneration rates were highest when cotyledons were cultured for 3 weeks in darkness on MS medium containing 2.5 per cent sucrose and a combination of IBA and thidiazuron. Regeneration rates for Flordaguard, Nemard and Memaguard were as high as 60 per cent, 33 per cent and 6 per cent respectively. Approximately 70 per cent of regenerated shoots produced rooted plants, thereby establishing a rapid and simple method of propagation.

Studies by Clemente *et al.* (1991) on *in vitro* propagation techniques of endangered and endemic Spanish *Artemisia granatensis* demonstrated that shoot proliferation was greatest in MS medium containing BA. Addition of auxins to the basal medium did not help in inducing roots. The species was successfully propagated from shoot tip explants. Microshoots were acclimatized in glass jar in peat for 4-5 weeks and produced good root systems.

Lubomski and Jerzy (1989) cultured stem internodes of pot carnation cv. Mini, Pinki on half strength MS medium having IAA and BA. Adventitious shoots were formed which regenerated maximum number of shoots on IAA supplemented medium.

The concept of *in vitro* micropropagation was simultaneously matched with somatic embryogenesis and diploid embroys developed from the explant in large numbers under the manipulated influence of growth regulators and each could develop directly into a complete plant, thereby producing large number of similar plants in a short span of time. Algination of the embryos led to the production of artificial seeds which could be stored at low temperatures for a long time.

Coutos-Thevenot *et al.* (1990) found that there was a strict relationship between the tetraploid state and the inability to produce somatic embryos. The embryogenic clones were all diploid. Somatic embryogenesis has been tried in a number of plant types, with success in few. Doorne *et al.* (1995) screened 16 cultivars of pea for the formation of somatic embryos and found it to be dependent on the genotype, culture conditions and explant source used. The development of somatic embryos to whole plants was not reliable and only 10 per cent plantlets were raised upto flowering, pod and viable seed production.

In *Camellia sinensis* (Wachira and Ogada, 1995) excised cotyledon explants produced somatic embryos without intermediate callus when cultured on MS medium containing 30 g sucrose/litre within 3 weeks of culture. The embryogenic competence was reduced by increasing concentration of kinetin BA and IBA. The embryos could grow to maturity without being subcultured within 6-8 weeks.

In *Crocus sativus*, Ahuja *et al.* (1994) somatic embryogenesis was initiated from shoot meristems in LS medium containing BA and NAA. The development of embryos was asynchronous and passed from globular stage to embryoids with bipolar regions. Mature embryos could be germinated on half strength MS medium with Gaz. Root tip squashes of plantlets regenerated had anormal chromosome number.

The successful development of somatic embryos in large numbers and their subsequent *in vitro* germination has led to the concept of artificial seeds. Embryogenic callus obtained from *in vitro* grown plantlets derived from meristems excised from lateral buds of horse radish was induced on solid MS medium supplemented with 2, 4-D or BA. Embryogenic cells obtained by transfer of this callus to liquid medium was fractionated and subcultured in growth regulator free medium. Somatic embryos obtained were first tested for their regenerative ability and then encapsulated in alginate Ca gel and placed on gelled MS medium without growth regulators. Encapsulated embryos could also regenerate successfully (Shigeta and Sato, 1994).

Repunte *et al.* (1995) obtained cell aggregates from horse radish hairy roots obtained by culturing roots derived from leaf discs with *Agrobacterium rhizogenes.* These cell aggregates were encapsulated in alginate gel and covered with thin paraffin layer to prevent regeneration of adventitious roots. These could be stored at 25° C for upto 60 days without losing their root regeneration potential. This is modified technique of artificial seed production because it does not require the development of somatic embryos. A still easier method has been reported by Kageyama *et al.* (1995) in *Pogostemon cablin* where isolated mesophyll protoplasts of *in vitro* grown patchouli leaves were encapsulated in alginate beads, approximately $2–3 \times 10^3$ protoplants per 25 µl bead. Successful colony formation was induced when protoplast beads were inoculated on a liquid medium supplemented with 10^{-6} M NAA and 10^{-5} MBA. Regeneration of plants from protoplast was achieved.

The application of tissue culture techniques to the production of useful secondary compounds has been recognised since the early 1950s. The significance of this technique has already been recognised in a large number of plants and can be applied for production of high amounts of vitamins, pigments, alkaloids, food flavours and useful metabolites (Collins and Watts 1983, Constabel and Vasil, 1987).

Rokem (1991) studied the formation of diosgenin, the main precursor for the manufacture of corticosteroids and steroid hormones in suspension cultures of *Dioscorea deltoidea.* Bioregulators were applied to inhibit metabolite pathways not necessary for the growth of cells in suspension thereby improving the diosgenin yield. Fungal cell walls added to the suspension enhanced the accumulation of diosgenin indicating that this product may be phytoalexin.

Some sesquiterpene lactones occuring in the Compositae are antibiotics, phytotoxins or cytotoxic. The production of sesquiterpene lactones in callus derived from sub-apical leaf explants of *Ambrosia tenuifolia* were studied by Goleniowski *et al.* (1990). These lactones were produced in greater amounts in callus than in whole plants growing under natural conditions. The effects of pH, sucrose concentration, oxygen pressure, light and darkness on callus growth and sesquiterpene production was studied a 2 per cent sucrose, pH 7 and illumination 12 W/m^2 was found to be optimum for its maximum production.

The production of catharanthine was found to increase in suspension cultures *Catharanthus roseus* grown at low light intensity. High light intensity was inhibitory to cell growth and also alkaloid production. (Park *et al.,* 1990) Yun *et al.* (1990) were able to optimise carotenold biosynthesis in carrot by controlling sugar concentration. A high sucrose concentration was associated with a low growth rate and final cell mass and cell size was small in cultures with low sucrose concentration. However, carotenoid biosynthesis was greatest with high sucrose levels in small rather than large cells. The best results were obtained by culturing cells initially with low sucrose to increase growth rate and then when cell mass had attained a certain level, raising the sucrose concentration and enhancing carotenoid biosynthesis.

Tori and Sakurai (1995) investigated the effects of added riboflavin and increased sucrose on cell growth and anthocyanin accumulation in suspension cultures of strawberry C.V. Srikenari petiole tissue. Riboflavin reduced cell growth markedly, however anthocyanin content using a medium with riboflavin at 4 mg/I was about thrice than that in the control media and the duration of culture was half that of control.

Somaclonal variation is one of the recent well recognised techniques of inducing novel type of variability under in vitro conditions and is assumed to take place due to epigenetic factors, chromosomal changes during culture, effect of medium constituents and also transposable nature of certain sequences.

Schwenkel and Grunewaldt (1990) grew rooted shoots of *Cyclamen persicum* from *in vitro* culture of eight diploid genotypes until flowering giving about 6000 plants. Variations in growth habit, flower shape and colour, chlorophyll pattern and leaf shape were noted. The frequency of altered plants ranged from 4.5 to 36.3 per cent and the type of alteration depended on original genotype. Giant variants were often tetraploid, some diploid but proved unstable following repeated clonal propagation. Dwarf forms were comparatively stable. Changes in flower colour and chlorophyll patterning occurred rarely and were assumed to be due to pre-existing genetic variation or unstability.

Raghunath and Priyadarshan (1993) regenerated four lines of plantlets of cardamon from axenic culture of juvenile shoot primordia and compared them with parental clone for variations after three years. Considerable variation was found in plant height and panicle branching character but not in number of tillers and leaves. Panicle size directly affects yield and therefore it was considered necessary that variants should be detected early so that they are not introduced in the micropropagation protocol. Florets and corollas of 15 cultivars and lines of chrysanthemum (*Pendranthema morifolium*) were cultured *in vitro* (Mizutani and Tanaka, 1994) and evaluated for regeneration and variation rates changes in characters such as flower colour, floret shape and floret number per flower were observed in plants regenerated from florets of 3 lines which were considered to be useful as new cultivars of Higo chrysanthemums.

Israeli *et al.* (1991) have given a qualitative description of somaclonal variants of banana with regard to plant stature, abnormal leaves, pseudostem pigmentation, persistence of flowers and split fingers. Dwarf fingers were most common while tall variants were rare. Another common variation was the appearance of thick and rubbery narrow leaves with variable pale green mottling.

Smith (1995) has considered the aspect of Begonias as a natural source of food colouring. Explants from various parts of in vitro cultured plantlets of *Begonia* c.v. *Rich mondensis* and *B. semperflorens* c.v. Vodka were studied for their ability to generate callus cultures with a view to in vitro production of anthocyanin. Callus proliferation was difficult to maintain and there was a strong tendency to revert to organogenesis. However, addition of thictiazuron to the culture medium at low concentration and the elimination of micronutrients including iron from the medium resulted in consistent formation of callus from all explant sources. Pigment production was then successfully induced in these cultures.

One of the most advanced techniques of gene manipulation in present days is through transformation which involves the transfer of a specific gene via a vector or other physical means to the recepient organism. *Petunia hybrida* plants (Lu *et al.*, 1989) transformed with an antisense chalcone synthase gene had white rather than blue flowers. Similar systems are being developed for carnation, chrysanthemum, rose and gerbera with the aim to modify flower colour and improve post harvest life of cut flowers. Transgenic carnation plants (*Dianthus caryophylius*) were obtained on kanamycin containing medium after cocultivation of stems with disarmed or aimed strains of *Agrobacterium tumefaciens* (Lu *et al.*, 1991).

Growth of *Atropa belladona* hairy roots in shake flasks was associated with synthesis of tropane alkaloids. The hairy roots were produced due to transformation with *Agrobacterium rhizogenes*. (Sharp and Doran, 1990). Sauerwein *et al.* (1991) trasformed *Hyoscyamus albus* with *Agrobacterium rhizogenes* leading to the production of hairy roots. A new piperidone alkaloid, hyalbidone was isolated from the hairy roots. Flores (1992) has considered plant roots as chemical factories on the basis of the biosynthesis of useful chemicals by them. The use of *in vitro* culture techniques, particularly the induction of hairy root cultures by *Agrobacterium rhizogenes* to increase the production of respective chemicals and the related manipulation of root metabolism are significant achievements.

Leaf pieces of 6 *Dendranthema morifolium* and one *D. indicum* genotypes were screened on a range of shoot regeneration medium. *D. indicum* had the highest response on basal medium supplemented with 0.2 mg IAA with 3-5 mg BA/litre. Transgenic plants of this genotype were generated using transformation mediated by the disarmed strain of *Agrobacterium tumefaciens* L BA 4404. This contained either pKIWI 110 or pGA 643 both of which contain the selectable marker gene NPT II and pKIWI 110 also contains GVS reporter gene. Leaf pieces inoculated with pKIWI 110 produced zones of blue cells two days after inoculation. Shoots from leaf pieces inoculated with pGA 643 were selected on kanamycin. PCR and Southern analysis of shoots that were able to root on kanamycin confirmed the presence of the N PT 11 gene in the plant genome. (Ledger *et al.*, 1991).

Agrobacterium rhizogenes transformed roots of *Nicotiana glauca* synthesise the alkaloids nicotine and anabasine at levels reflecting the parent plants. Media composition, strength and pH were evaluated with respect to biomass yield and productivity (Green *et al.*, 1992). Full strength Gamborg's B5 medium proved the best for biomass yield while half strength or low salt medium enhanced alkaloid accumulation. High nitrate concentrations enhanced media alkaloid levels at the end of the growth phase. The pH of medium is also important. Transformed roots could grow between pH 3 and 9 but root biomass is favoured by an increase in medium alkalinity while released alkaloid is favoured by milky acidic pH. Transformed roots release a portion of their secondary metabolites into the growth medium. By continually removing root products any feedback inhibition on enzymatic reaction is reduced as are the toxic plants resulting from product accumulation.

It can thus be seen that during the recent years much emphasis is being laid on the application of the modern established techniques of multiplication and induction of genetic variability to a vast range of horticultural crops both in India and abroad. The conventional processes need to be modified greatly in almost all the crops for targetted increase in yield, improvement of quality and production of secondary metabolites.

As is with all other crop species–cereals and cash crops, the horticultural crops are also undergoing a period of renaissance where techniques of biotechnology are being tried and tested for judicious application in order to make the plants capable of meeting the demands of the ever growing population.

References

Ahuja, A., Koul, S., Rain, G. and Kaul, B.L., 1994. Somatic embryogenesis and regeneration of plantlets in saffron (*Crocussativus* L). *Ind. J. Exp. Biology*, 32(2): 135–140.

Christopher, T. and Rajan, M.V., 1994. *In vitro* clonal propagation of *Capsicum* spp. *Plant cell, Tissue and Organ Culture*, 38(1): 25–29.

Clemente, M., Contreras, P., Susin, J. and Pliego-Alfaro, F., 1991. Micropropagation of *Artemisia granatensis. Hort. Science*, 26(4): 420.

Collins, H.A. and Walts, M., 1983. Flavor production in culture. In: *Handbook of Plant Cell Culture Vol. 1: Techniques for Propagation and Breeding*, (Eds.) D.A. Evans *et al.* MacMillan Publishing Co., N.Y., pp. 729–747.

Constabel, F. and Vasil, I.K. (Eds.), 1987. *Cell Culture and Somatic Cell Genetics of Plants, Vol. 4: Cell Culture in Photochemistry*. Academic Press Inc., San Diego, pp. 1–314.

Coutos-Therenot, P., Jouanneau, J.P., Brown, S., Petiard, V. and Guern, J., 1990. Embryogenic and non-embryogenic cellines of *Daucus carota* cloned from meristematic cell clusters: Relation with cell ploidy determined by flow cytometry. *Plant Cell Reports*, (10): 605–608.

Doorne, L.E. Van, Marshall, G. and Kirkwood, R.C., 1995. Somatic embryogenes is in pea (*Pisum sativum* L): Effect of explant, genotype and culture conditions. *Annals of Applied Biology*, 126(1): 169–179.

El Farash, E.M., Abdalla, H.I., Taghain, A.S. and Ahmed, Mol, 1993. Genotype, explant age and explant type as effecting callus and shoot regeneration in tomato. *Asian Journal of Agricultural Sciences*, 24(3): 3–14.

Flores, H.E., 1992. Plant roots as chemical factories. *Chemistry and Industry*, 10: 374–377.

Frey, L. and Janick, J., 1991. Organogenesis in carnation. *Journal of the American Society for Horticultural Science*, 116(6): 1108–1112.

Geslot, A., Connault, C., Merlin, G., El Maataqui, M. and Neville, P., 1989. *In vitro* multiplication of *Mentha* combining apex culture and microcuttings. *Bulletindeia Societe Botaniquede france Letters Botaniques*, 136(1): 31–38.

Goleniowski, M.B., Silva, G.L. and Trippi, V.S., 1990. Sesquiterpene lactone production in callus cultures of *Ambrosia tenuifolia*, 29(9): 2889–2891.

Green, K.D., Thomas, N.H. and Callow, J.A., 1992. Product enhancement and recovery from transformed root cultures of *Nicotiana giauca*. *Biotechnology and Bioengineering*, 39(2): 195–202.

Israeli, Y., Reuveni, O. and Lahav, E., 1991. Qualitative aspects of somaclonal variations in banana propagation by *in vitro* techniques. *Scientia Horticulturae*, 48(1–2): 71–88.

Kageyarna, Y., Honda, Y. and Sigimiva, Y., 1995. Plant regeneration from Petchouli retoplasts encapsulated in alginate beads. *Plant Cell Tissue and Organ Culture*, 41(1): 65–70.

Laszlo, M. and Berci, I., 1990. The effect of cultural conditions on the alkaloid production of *Catharanthus roseus* L. (G.Don) tissue cultures. *Napjaink Biotechnologiaja*, 26: 30–34.

Ledger, S.E., Deroles, S.C. and Given, N.K., 1991. Regeneration and *Agrobacterium* mediated transformation of chrysanthemum. *Plant Cell Reports*, 10(4): 195–199.

Lu, C.Y., Nugent, G., Wardley Richardson, T., Chandler, S.F., Young, R. and Dalling, M.J., 1991. *Agrobacterium* mediated transformation of carnation (*Dianthus caryophyllus* L). *Biotechnology*, 9(9): 86–868.

Lu, C.Y., Wardley, T., Gilmour, M., Nugent, C. and Mirabile, P., 1989. Genetic engineering of flowers at Calgene Pacific. *Australian J. of Biotechnology*, 3(4): 285–287.

Lubomski, M. and Jerzy, M., 1989. *In vitro* propagation of pot carnation from stem internodes. *Acta Horticulturae*, 25: 235–240.

Mizutani, T. and Tanaka, T., 1994. Study on the floret culture of Higo chrysanthemum. *Proc. of Faculty of Agriculture*, Kyushu Tokai University, 13: 9–14.

Park, H.H., Choi, S.K., Kang, K, and Leely, 1990. Enhancement of producing catharanthine by suspension growth of *Catharanthus roseus*. *Biotechnology Letters*, 12(8): 603–608.

Pooler, M.R. and Scorza, R., 1995. Regeneration of peach (*Primus persica* L. Batsch) rootstock cultivars from cotyledons of mature stored seed. *Hort Science*, 30(2): 355–356.

Raghunath, B.R. and Priyadarshan, P.M., 1993. Somaclonal variation in cardamon (*Elletaria cardamomum*) derived from axenic culture of juvenile shoot rimordia. *Acta. Horticulturae*, 330: 235–242.

Raut, R.S. and Lokhande, V.L., 1989. Propagation of plantain through meristem culture. *Annals of Plant Physiology*, 3(2): 256–260.

Repunte, V.P., Taya, M. and Tone, S., 1995. Preparation of artificial seeds using cell aggregates from horse radish hairy roots encapsulated in alginate gel with paraffin coat. *Journal of Fermentation and Bioengineering*, 79(1): 83–86.

Roken, J.S., 1991. Methods to increase formation of secondary metabolites in plant cell suspension cultures. *Israeli Journal of Botany*, 40(3): 264.

Sauerwein, M., Ishimaru, K. and Skinomura, K., 1991. A piperidone alkaloid from *Hypcyamus albus* roots transformed with *Agrobacterium rhizogenes*. *Phytochemistry*, 30(9): 2977–2978.

Schwenkel, H.G. and Grunewaldt, J., 1990. Somaclonal variation in *Cyclamen persicum* Mill. after *in vitro* mass propagation. *Eucarpia*, 62–67.

Seabrook, J.E.A., 1994. *In vitro* clonal propagation of *Capsicum* spp. *Plant Cell, Tissue and Organ Culture*, 38(1): 25–29.

Seyham, S. and Ozzambak, E., 1994. Shoot multiplication of some Olive (*Olea europaea* L) Cultivars. *Acta Horticulturae*, 30: 35–38.

Sharp, J.M. and Doran, P.M., 1990. Characteristics of growth and tropane alkaloid synthesis in *Atropa belladona* roots transformed by *Agrobacterium rhizogenes*. *J. Biotech.*, 16(3–4): 171–184.

Shigeta, J. and Sato, K., 1994. Plant regeneration and encapsulation of somatic embryos of horse radish. *Plant Science*, 102(1): 109–115.

Silvertand, B., Jacobzen, E., Mazereeuw, J., Lavry Sen, P. and Harten A. Van, 1995. Efficient in vitro regeneration of leek (*Allium ampeloprasum* L.) via flower stalk segments. *Plant Cell Reports*, 14(7): 423–427.

Smith, M.A.L., 1995. Begonias as a natural source of food clouring. *Begohian*, 62: 7–10.

Swamy, M.S., Christopher, T. and Subhash, K., 1988. Multiple shoot formation in embryo culture of *Solanum melongena*. *Current Science*, 57(4): 197–198.

Tori, T. and Sakurai, M., 1995. Effects of riboflavin and increased sucrose on anthocyanin production in suspended strawberry cell cultures. *Plant Science*, 110(1): 147–153.

Turakhia, D.V. and Kulkarni, A.R., 1988. *In vitro* regeneration from leaf explants of *Ladebouria hyacinthiana* Roth. (*Scilla indica*). *Current Science*, 57(4): 214–216.

Wachira, F. and Ogada, J., 1995. *In vitro* regeneration of *Camellia sinensis* (L) O. Kuntze by somatic embryogenesis. *Plant Coll Reports*, 14(7): 463–466.

Yun, J.W., Kim, J.H. and Yoo, Y.J., 1990. Optimizations of carotenoid biosynthesis by controlling sucrose concentrations. *Biotechnology Letters*, 12(12): 905–910.

Zirari, A. and Lionakis, S.M., 1994. Effect of cultivar, explant type etiolation pretreatment and the age of plant material on the *in vitro* regeneration ability of avocado (*Persea americana*). Acta *Horticulture*, 365: 69–75.

Chapter 8

Tissue Culture Techniques for Developing Weevil Resistance in Sweet Potato

Harsh Kumar and S.C. Choudhary

*Department of Genetics, Rajendra Agricultural University,
Pusa, Samastipur – 843 125, Bihar*

ABSTRACT

Weevil is the worst enemy of sweet potato. It is difficult to control by conventional pest management system. In built resistance do the weevil to exist in some germplasm which can be transferred to the desired genotypes through hybridization. However, the hybridization in sweet potato is limited by shyness of flowering, the presence of incompatibility and fertility barrier, poor capsule and seed setting, which limit crossibility for the transfer of weevil resistance from highly resistant genotypes to the ruling cultivars. Thus, a view to transfer weevil resistance, four genotypes X-145, 86-X-17 (highly resistant), X-127 (susceptible) and Mix-OP-93-I (highly susceptible), which showed profuse synchronus flowering, high pollen fertility and some fruit setting, were hybridized in all possible combinations resulting into a very low fruit setting ranging from 0.50 to 2.0 per cent. The number of seeds per capsule were low and most of the seeds were non-viable. The hybrid embryos from the non-viable seeds were rescued using tissue culture method, which also resulted into their micropropagation. Thus, the work showed the possibility of utilizing tissue culture techniques for developing weevil resistance in sweet potato.

Introduction

Sweet potato (*Ipomoea batatas* L.) is an important dietary staple for the tropical countries with high nutritional and calorific values. It can be grown with minimum inputs and can provide a very high return. Such an important crop is most severely damaged by sweet potato weevil (*Cylas formicarius* Fabricius), which is difficult to control by conventional pest management system due to its cryptic Feeding habit and nocturnal activity. Thus, search for an inbuilt resistance against the weevil in the sweet potato genotypes and their transfer to commercial cultivar is required (Talekar, 1987). The progress in the area of developing host plant resistance to sweet potato weevil has been slow. In spite of over 50 years of research, no sweet potato cultivar has been found in nature or developed to be totally immune to *Cylas formicarius* (Barlow and Rolston, 1981; Mullen *et al.*, 1980; Rolston *et al.*, 1979). Jansson *et al.* (1991) also reported that no cultivar with high level of resistance is available. However, low level of resistance to weevil do exist in different germplasm pools of sweet potato. Choudhary *et al.* (1997) found some sweet potato germplasm resistant to weevil. Presence of cross incompatibility between some genotypes along with shyness in flowering habit prohibit the transfer of weevil resistance and limit the breeding of sweet potato for resistance to weevil through conventional method. To overcome these difficulties tissue culture techniques may be utilized for developing weevil resistance in sweet potato. The present investigation is an effort in this direction.

Materials and Methods

Selection of Parents

Suitable genotypes were selected among 703 germplasm on the basis of weevil resistance level, flowering habit, flowering period, pollen fertility (per cent) and open fruit setting. These selected genotypes were subjected to hybridization.

Hybridization

Selected genotypes were used in hybridization programme. Hybridization was done as per details given below:

Emasculation

In evening, one day before opening of the flower, the buds were carefully forced open and five stamens were removed by forceps. Then emasculated buds were bagged and left for pollination.

Pollination

Emasculated flowers were debagged and pollinated by appropriate pollen. After pollination the flower was rebagged. Mostly the pollination was done from 2.00 to 6.00 a.m. under artificial light in the month of November to January in 1994-95 and 1995-96. The frequency of successful hybridization as evident by number of fruit set was calculated. The time (in days) required for ripening of fruit and number of seeds per fruit were recorded. The number of hybrid seeds were very low and most of the seeds were nonviable, so embryo rescue was used to raise the hybrids.

Rescue of Hybrid and its Micropropagation

Mature hybrid seeds were surface sterilized and placed over sterilized moist filter papers under aseptic conditions. To select proper media for rescue of hybrid embryo, initially open pollinated seeds were cultured on MS (Murashige and Skoog, 1962) basal medium supplemented with different combinations and concentrations of NAA (α-Naphthalene acetic acid), IBA (Indole-3 butyric acid), BAP (6-Benzyl amino purine) and KIN (Kinetin) and finally four media were selected namely MS + NAA (0.5 mg L^{-1}) + BAP (0.5 mg L^{-1}) for seed culture, MS + NAA (0.1 mg L^{-1}) + BAP (1.0 mg L^{-1}) for rescue of hybrid embryo, MS + NAA (0.1 mg L^{-1}) + BAP (5.0 mg L^{-1}) for multiplication of hybrids and $1/2$ MS + IBA (0.2 mg L^{-1}) for rooting of the hybrid shoots. After successful rooting, plantlets were transferred into pots and subjected to hardening. Hardened plants were successfully transferred into the field condition for their evaluation and utilization in development of weevil resistance.

Results and Discussion

To select weevil resistance in sweet potato germplasm and to transfer this resistance into existing cultivars, the experiment were conducted with respect to selection of parents, hybridization, rescue of hybrid embryo and micropropagation of rescued hybrid.

Selection of Parents

Clones of suitable genotypes namely X-145 (highly resistant), X-127 (susceptible), 86-X-17 (highly resistant) and Mix-OP-93-1 (highly susceptible) were selected on the basis of flowering habit, flowering period, pollen fertility (per cent) and open fruit setting (Table 8.1). These genotypes showed moderate to very profuse synchronous flowering, pollen fertility more than 80 per cent and very profuse open fruit setting during November to January.

Table 8.1: Flowering Behaviour and Open Seed Formation in Sweet Potato

Sl.No.	Genotypes	Flower Colour	Flowering Habit	Flowering Period	Pollen Fertility (%)	Open Fruit formation	Fruit Maturity Period (Days)
1.	X-145	Pale purple limb with purple throat	+++	15 Nov.–15 Jan.	80.00	+	50–65
2.	X-127	White	+++++	Nov.–Jan.	95.00	+++	50–60
3.	86-X-17	Pink	++++	15Nov.–20Jan.	93.00	++	55–65
4.	Mix-OP-93-1	-do-	+++	Nov.–Jan.	96.00	++	50–60

+: Scarce; ++: Sparse; +++: Moderate; ++++: Profuse; +++++: Very Profuse.

The selected genotypes for hybridization experiment represented highly resistant, susceptible and highly susceptible types. During the selection of these genotypes the flowering habit, flowering period, pollen fertility (per cent) and open fruit setting were considered because of the general shyness of flowering and poor fertility in the sweet potato germplasm (Martin, 1967; Srinivasan, 1977; Vimala, 1989). The selected genotypes showed synchronous flowering, high pollen fertility and some fruit setting indicating their suitability for the breeding programme.

Hybridization

Selected parents were crossed in different possible combinations (Table 8.2). Very poor fruit setting were observed ranging from 0.50 to 2.00 per cent with the maximum in the cross X-127 × Mix-OP-93-1. The total seeds formed from each cross were only a few ranging from 1 to 8. As the number of seeds formed were only a few and the hybrid seeds were mostly nonviable, embryo culture was done to rescue the hybrid. The poor fruit setting during hybridization may be due to cross incompatibility. The different factors including incompatibility that affect fruit set in sweet potato have been discussed (Fujise, 1964; Martin, 1965, 1967, 1982; Srinivasan, 1977; Srinivasan and Vimala, 1981; Vimala, 1989). These problem limit the development of desired genotypes including those resistant to weevil through the breeding programme. The best cross during the investigation was found to be X-127 × Mix-OP-93-1.

Table 8.2: Results of Hybridization in Sweet Potato

Sl.No.	Crosses	No. of Flower Pollinated	No. of Well Developed Fruit	% of Fruit Formation	Fruit Maturity Period (Days)	Total Seed formation
1.	X-145 × Mix-OP-93-I	205	3	1.46	50-65	3
2.	X-145 × 86-X-17	220	4	1.82	-do-	7
3.	X-145 × X-127	200	2	1.00	-do-	3
4.	X-127 × Mix-OP-93-I	250	5	2.00	-do-	8
5.	X-127 × X-145	210	2	0.95	-do-	5
6.	X-127 × 86-X-17	212	4	1.89	-do-	6
7.	86-X-17 × Mix-OP-93-1	195	1	0.50	-do-	2
8.	86-X-17 × X-145	217	3	1.38	-do-	5
9.	86-X-17 × X-127	213	2	0.94	-do-	4
10.	Mix-OP-93-1 × X-127	200	1	0.50	-do-	1
11.	Mix-OP-93-1 × X-127	298	3	1.01	-do-	6
12.	Mix-OP-93-1 × X86-X-17	204	4	1.96	-do-	8

The hybrid seeds did not germinate when cultured on the medium MS + NAA (0.5 mg L^{-1}) + BAP (0.5 mg L^{-1}) but swelled consipicuously. After 4-5 weeks of seed culture, the embryos were dissected out and cultured on the embryo culture medium MS + NAA (0.1 mg L^{-1}) + BAP (1.0 mg L^{-1}). Only shoots were formed without any root development. Among the hybrids, X-127 × X-145 showed the best response with 80 per cent embryos surviving and 60 per cent forming hybrid shoots (Table 8.3). Hybrids Mix-OP-93-1 × X-127 and Mix-OP-93-1 × 86-X-17 showed the least response.

**Table 8.3: Hybrid Embryo Culture on MS + NAA (0.1 mg L⁻¹) +
in Sweet Potato BAP (1.0 mg L⁻¹)**

Hybrids	Per cent Embryo Showing Response	Per cent Embryo that Formed Shoots
X-145 × 86-X-17	40.0	20.0
X-127 × Mix-OP-93-1	40.0	40.0
X-127 × X-145	80.0	60.0
X-127 × 86-X-17	60.0	40.0
86-X-17 × X-145	60.0	40.0
Mix-OP-93-1 × X-127	40.0	20.0
Mix-OP-93-1 × 86-X-17	20.0	20.0

* Results of 5 seed culture/hybrid.

Hybrid production through sexual crosses may be prevented by incompatibility barriers. Such barriers may be overcome by tissue culture techniques. Specially in cases when fertilization is successful but embryo fail to develop, the culture of such embryo can rescue the hybrid plant facilitating the transfer of desired traits among the genotypes (Stalkar, 1980). Embryo rescue has been successful in sweet potato. However, some crosses showed better success than others indicating a role of genotype as found in other tissue culture responses in sweet potato (Choudhary *et al.*, 1996; Choudhary and Kumar, 2000). Embryo culture has been used to develop plants from non-viable seeds *of* another tuber crop cassava (Ng, 1992). In sweet potato, however, ovule culture have been used to raise the hybrid (Wang *et al.*, 1998).

Since, the number of developed hybrid shoots was very low, their nodal stem and apical shoots were subcultured into the medium MS + NAA (0.1 mg L⁻¹) + BAP (5.0 mg L⁻¹) which resulted into multiple shoot formation. Among the different hybrids, X-127 × X-145 showed the best response for percentage cultures showing multiple shoot formation as well as number of shoots per culture followed by X-127 × 86-X-17 and X-127 × Mix-OP-93-1 (Table 8.4).

**Table 8.4: *In vitro* Shoot Multiplication through Nodal Stem Culture on MS + NAA
(0.1 mg L⁻¹) + BAP (5.0 mg L⁻¹) in Sweet Potato**

Hybrids	Per cent Culture Showing Multiple Shoot Formation	No. of Shoot per Culture
X-145 × 86-X-17	33.33	3.00
X-127 × Mix-OP-93-I	40.00	3.50
X-127 × X-145	50.00	3.75
X-127 × 86-X-17	50.00	2.50
86-X-17 × X-145	25.00	3.00
Mix-OP-93-1 × X-127	0.00	0.00
Mix-OP-93-1 × 86-X-17	0.00	0.00

Cytokinins are generally considered essential for *in vitro* multiple shoot formation (Bhojwani, 1980). Among cytokinins, BAP is the most widely used and the most effective. Wang *et al.* (1998) have also found BAP essential for multiple shoot formation in ovule culture of sweet potato. For multiple shoot formation also, genotype played an important role.

The *in vitro* developed multiple shoots were transferred to the rooting medium 1/2 MS + IBA (0.2 mg L⁻¹). Depending on genotypes, low to profuse rooting were observed (Table 8.5). IBA is a known potent root inducer and has been used for rooting of *in vitro* developed shoots during micropropagation of many plants. However, Xue (1988) has achieved root formation on medium 1/2 MS + IAA on *in vitro* developed shoots of sweet potato.

Table 8.5: Root Development at the Base of *in vitro* Formed Shoot on 1/2 MS + IBA (0.2 mg L⁻¹) in Sweet Potato

Hybrids	Root Development
X-145 × 86-X-17	+
X-127 × Mix-OP-93-I	++
X-127 × X-145	+++
X-127 × 86-X-17	++
86-X-17 × X-145	++

+: Low rooting; ++: Good rooting; +++: Profuse rooting.

The rooted plantlets were progressively acclimatized and hardened for their final field transfer. The survival during hardening was low and showed genotype specificity. Only the plantlets of hybrid X-127 × X-145) and X-127 × Mix-OP-93-1 survived and were transferred to the field condition. The plantlets of other genotypes generally showed decay at the shoot base and finally died Field transfer of *in vitro* developed plantlets is generally a difficult process with very few successes. In another tuber plants *Dioscorea esculenta* also the *in vitro* developed plantlets died during field transfer due to root decay (Nair and Chandra Babu, 1996).

Thus, the present work resulted into rescue of hybrids and their propagation through tissue culture and the evaluation of these hybrid plants may facilitate the transfer of weevil resistance into desired genotypes of sweet potato.

References

Barlow, T. and Rolston, L.H., 1981. Type of host plant resistance to the sweet, potato weevil found in sweet potato roots. *J. Kansas Entomol. Soc.*, 54: 649–657.

Bhojwani, S.S., 1980. Micropropagation method for a hybrid willow (*Salix matsudana* × alba NZ–1002). *New Zealand J. Bot.*, 18: 209–214.

Choudhary, S.C. and Kumar, H., 2000. Genotypic effect on callusing from leaf explant of sweet potato. *RAU. J. Res.* (Accepted).

Choudhary, S.C., Kumar, H., Nasar, S.K.T. and Verma, V.S. 1996. Callusing response of sweet potato, *Ipomoea batatas* (L.) Lam, genotypes. In: *Tropical Tuber Crops: Problems, Prospects and Future Strategies* (Eds.) G.T. Kurup, M.S. Palaniswami, V.P. Potty, G. Padmaja, S. Kabeerathumma and S.V. Pillai. Science Publishers, Labanon, New Hampshire, U.S.A., pp. 43–48.

Choudhary, S.C., Kumar, H., Nasar, S.K.T., Verma, V.S., Kumar, M. and Faruqui, O.R., 1997. Search for weevil resistance in sweet potato germplasm. In: *Proc. 3rd Agril. Sci. Cong.*, Vol. 2 (Eds.) V. Beri, V.K. Dilawari and M.S. Bajwa. P.A.U., Ludhiana, India, p. 208.

Fujise, K., 1964. Studies on flowering, fruit setting and self and cross incompatibility in sweet potato varieties. *Bull. Kyushu Agric. Exp. Sta.*, 9: 123–146.

Jansson, R.K., Raman, K.V. and Malamud, O.S., 1991. Sweet potato pest management: Future outlook. In: *Sweet Potato Pest Management: A Global Perspective*, (Eds.) R.K. Jansson and K.V. Raman. Oxford and IBH Pub. Co. Pvt. Ltd., New Delhi, pp. 429–437.

Martin, F.W., 1965. Incompatibility in the sweet potato: A review. *Econ. Bot.*, 19: 406–415.

Martin, F.W., 1967. The sterility incompatibility complex of the sweet potato. *Proc. Int. Symp. on Root Crops* (Trinidad), 1: 1–15

Martin, F.W., 1982. Analysis of the incompatibility and sterility of sweet potato. *Proc. First Int. Symp. on Sweet Potato*, 11: 275–296.

Mullen, M.A., Jones, A., Arbogast, R.T., Schalk, J.M., Paterson, D.R., Boswell, T.E. and Earhart, D.R., 1980. Field selection of sweet potato lines and cultivars for resistance to sweet potato weevil. *J. Econ. Entomol.*, 73: 288–290.

Murshige, T. and Skoog, F., 1962. A revised medium for rapid growth and bioassay with tobacco tissue cultures. *Physiol. Pl.*, 15: 473–476.

Nair, N.G. and Chandra Babu, S., 1996. *In vitro* production and micropropagation of three species of edible yams. In: *Tropical Tuber Crops: Problems, Prospects and Future Strategies* (Eds.) G.T. Kurup, M.S. Palaniswami, V.P. Potty, G. Padmaja, S. Kabeerathumma and S.V. Pillai. Science Publishers, Labanon, New Hampshire, U.S.A., pp. 55–60.

Ng, S.Y.C., 1992. Tissue culture of root and tuber crops at IITA. In: *Biotechnology: Enhancing Research on Tropical Crops in Africa*, (Eds.) G. Thottappilly, L.M. Monti, D.R. Mohan Raj and A.W. Moore. IITA, Ibadan, Nigeria, pp. 135–141.

Rolston, L.H., Barlow, T., Hernandez, T., Nilakhe, S. and Jones, A., 1979. Field evaluation of breeding lines and cultivars of sweet potato for resistance to the sweet potato *weevil. Hort. Sci.*, 14: 634–635.

Srinivasan, G., 1977. Factors influencing fruit set in sweet potato. *J. Root Crops*, 3: 53–57.

Srinivasan, G. and Vimala, B., 1981. Studies on incompatibility and fruit, set in different cultivars of sweet potato. *J. Root Crops*, 7: 21–24.

Stalkar, H.T., 1980. Utilization of wild species for crop improvement. *Adv. Agron.*, 33: 111–147.

Talekar, N.S., 1987. Resistance in sweet potato to sweet potato weevil. *Insect Sci. Applic.*, 8: 819–S23.

Vimala, B., 1989. Fertility and incompatibility in sweet potato (*Ipomoea batatas* L.). *Ann. Agric. Res.*, 10: 109–114.

Wang, Jia Xu, Lu Shu Yun, Zhou Hai Ying and Liu Qing Chang, 1998. The second report of studies on overcoming the interspecific incompatibility and hybrid abortion between series A and B in the section *Batatas* of the genus *Ipomoea*. *Ada Agronomica Sinica*, 24: 139–146.

Xue, Q.H., 1988. Organ formation and plantlet regeneration from unfertilized ovule callus of *Ipomoea* species. *Jiangsu J. Agril. Sci.*, 4: 56–62.

Chapter 9

Improvement of Protein Quality and Quantity in French Bean

P.R. Prasad

Mutagenesis and Cytogenetics Laboratory,
Department of Botany, R.S. College, Muzaffarpur – 842 001, Bihar

ABSTRACT

Variability was induced in salt soluble protein and methionine content in *Phaseolus vulgaris* after alkanesulfonates (EMS, MES and MMS) and antibiotics (Ampicillin, Neomycin, Furadentin) treatments. A significant shit in mean, range and variance in both the traits was observed in M_2. The significant positive correlation between these two traits was obtained after treatment suggested that restructuring of plants having high-protein and methionine content in French bean is possible through mutation breeding.

Introduction

Although, Legumes contribute 18 per cent of world requirement of protein (Jalil and Tahir, 1973), its protein quality is inferior because of low amount of sulphur-containing amino acids (Zimmermann and Levy, 1962; Bandemer and Evans; 1963; Abu-Shakra *et al.*, 1970; Prasad and Haider, 1985). The stored seed protiens in bean seed arc globulines (Pant and Tulsiani, 1969). The globulines are salt soluble and characterized by low contents of sulpher-containing amino acids. Naismith (1955) separated 4 components *viz.* 2, 7, 11 and 15^s from soyabean globuline. The 7^s component of globuline has low methionine content. Thus qualitative improvement of bean seed protein is possible if synthesis of the 7^s component is suppressed genetically (Nelson, 1969). However, to date, no substantial work has been done in this direction. Recognising these facts and uniqueness of induced mutation, the present investigation was carried out with a view to isolate mutants having high protein associated with balanced methionine contents in *Phaseolus vulgaris*.

Materials and Methods

Hundred dry and dormant seeds each of *Phaseolus vulgaris* was treated with 0.25 per cent and 0.50 per cent freshly prepared solution of alkylating agents [EMS (Ethyl Methane Sulfonate), MES (Methyl Ethane Sulfonate), MMS (Methyl Methane Sulfonate)] and antibiotics [AMP (Ampicillin), NEO (Neomycin) FUR (Furadentin)] for four hours under laboratory conditions with constant shaking. Hundred seeds were soaked in distilled water for the period chemical treatments were given. Seeds were thoroughly washed with distilled water to remove the traces of chemicals. All mutagenized and non-mutagenized seeds were sown in the field. The M_1 plants were individually harvested. Random sample of seeds from each fertility group (0-5, 6-10,- - - -96-100) with a total of hundred seeds per treatment were sown to raise M_2 population. Seeds of M_2 population were also collected plant wise.

Five seeds from each M_2 Plants were grinded separately in 'Willy mill' to fine powder of 60 mesh, 500 mg of grinded sample was dissolved in 50 ml of 0.5M NaCl solution. The content after shaking over an electric shaker for an hour was kept overnight for extraction. Extract was centrifuged at 6000 rpm for 30 minutes. The supernatant was then transferred to 50 ml volumetric flask and then volume was made up by adding 0.5M NaCl solution. The extract thus obtained was used for estimation of salt soluble fraction of protein after Lowry *et al.* (1951).

The methionine content of M_2 seeds were estimated plantwise following MacCarthy and Sullivan, 1941.

Correlation between salt soluble protein and methionine was calculated following formula given by Karl Pearson (Panse Sukhatme, 1967).

Results

Mean salt soluble protein and total methionine content expressed as mg/gm of seed on 13 per cent moisture basis is given along with range and variance in Table 9.1. A unidirectional shift in mean of salt soluble protein towards positive direction was noticed at M_2 after both the group of chemicals. The maximum increase in mean as compared to control (30.22±0.36) was noticed in 0.25 MES (38.02±0.74) among alkylating agents and 0.25 per cent FUR (41.56±0.90) among antibiotics treatments (Table 9.1).

However, a bidirectional shift of mean towards positive and negative direction was recorded for methionine content (Table 9.1). Highest and lowest mean in relation to control (2.91±0.06) was noticed after 0.50 per cent AMP (3.13±0.07) and 0.50 per cent FUR (1.84±0.08) respectively.

An appreciable increase in range and variance (Table 9.1) as compared to control for both protein and methionine content was recorded after both the concentrations of alkylating agents and antibiotics.

Table 9.1: Effect of Alkylating Agents and Antibiotics on Salt Soluble Protein Fraction and Total Methionine Contents (mg/gm seed sample) in M_2 Generation

Treatments	Salt Soluble Protein mg/gm Seed Sample			Methionine mg/gm Seed Sample			Correlation Co-efficient (r)
	Mean mg/gm±S.E	Range	Variance	Mean mg/gm±S.E	Range	Variance	
Control	30.22±0.36	22.89–34.88	6.57	2.91±0.06	2.33–3.33	0.04	0.9279 N.S
0.25 per cent EMS	34.58±1.28	27.25–46.87	6.7	3.02±0.13	2.00–4.33	0.67	0.6354 N.S
0.50 per cent EMS	36.74±0.77	29.43–44.69	16.98	2.97±0.12	2.16–3.66	0.48	0.2566**
0.25 per cent MES	38.02±0.74	26.16–46.87	20.2	2.93±0.07	1.66–3.79	0.23	0.2968**
0.50 per cent MES	30.44±1.23	26.16–34.88	36.64	2.39±0.10	1.82–3.16	0.26	0.0514**
0.25 per cent MMS	37.81±0.75	28.32–46.87	15.94	2.57±0.11	1.66–3.69	0.36	0.01629**
0.50 per cent MMS	NO SURVIVAL						
0.25 per cent AMP	34.20±0.98	26.16–45.78	28.15	2.47±0.15	1.33–3.66	0.72	0.9921 N.S
0.50 per cent AMP	38.97±0.88	31.61–49.03	25.02	3.13±0.07	2.49–3.79	0.16	0.6935 N.S
0.25 per cent NEO	32.87±0.65	23.98–39.24	14.49	1.93±0.07	1.00–2.69	0.18	0.1540**
0.50 per cent NEO	32.06±0.73	23.98–39.25	13.44	1.98±0.08	1.66–2.82	0.2	0.7298 N.S
0.25 per cent FUR	41.56±0.90	33.79–51.22	36.97	2.42±0.09	1.49–3.16	0.34	0.7298 N.S
0.50 per cent FUR	32.07±0.63	26.16–39.24	10.84	1.84±0.08	1.00–3.00	0.19	0.0697 N.S

N.S: Non-significant.

** Significant at 1 per cent level.

Correlation coefficient of salt soluble protein and methionine content (Table 9.1) indicated non-significant positive correlation in control, whereas this association was found to be highly significant in most of the treated populations at M_2.

Discussion

Subsequent to the discovery of Opaque-2 in maize (Mertz *et al.*, 1964) and Hiproly in barley (Hagberg and Karlsson, 1969), an intensive search for high-protein quality and quantity has been carried out, particularly in cereals. Prasad and Prasad (1981a) reported variation in methionine content after differential and combined treatments of alkane sulfonates and antibiotics in *Phaseolus vulgaris*. A little genetic variability in salt soluble fraction of protein and desirable amino acid contents has been reported by Schons (1970) in barley. Considerable variation in crude protein and essential amino acids contents in seeds of 22 *Phaseolus* sps. has been observed by Baldi and Salamini (1973). They concluded that the genus *Phaseolus* possesses a theoretical potential for the synthesis of reserve protein, with a balanced sulphur-containing amino acids content. But the practical realization is conditioned by the existence of incompatibility barriers which restrict crossing among *Phaseolus* sps.

In the present investigation an appreciable increase in mean range and variance for protein and methionine contents have been generated as a result of treatments with alkylating agents and antibiotics. Although, few treatments indicated almost equal or decreased mean methionine content.

An interesting part of the present investigation is that, though the correlation between globuline fraction and methionine content was not significant in control but this association was found to be highly significant after 0.25 per cent MES, MMS, NEO and 0.50 per cent EMS and MES is indicative of the fact that qualitative and quantitative improvement of protein could be obtained only after ascertaining an optimum dose of mutagens.

Increased mean, range, variance and significant positive correlation observed after mutagenetic treatments, provide an ample scope for selection of desirable plant having high protein and methionine content. This further raises the hope that restructuring of plant with balanced protein is possible through mutation breeding.

Undoubtedly, it is only important from an academic and scientific point of view, but it cannot be denied that antibiotics-a putative group of mutagens, too induced variability similar to that of alkylating agents-a well known group of mutagens. This supports an earlier observation (Prasad and Prasad, 1981a and 1981b) that antibiotics function like potential mutagens.

References

Abu-Shakra, S., Nirza, S. and Tannous, R., 1970. Chemical composition and amino acid content of chickpea seeds at different stages of development. *J. Sci. Food Agric.*, 21: 91–93.

Baldi, G. And Salamini, F., 1973. The Variability of essential amino acid content in seeds of 22 Phaseolus Species. *Theoretical and Applied Genetics*, 43: 75–80.

Bandmer, S.L. and Evans, R.J., 1973. The amino acid composition of some seeds. *J. Agric Food Chem.*, 11: 134–137.

Hagberg, A. and Karlsson, K.E., 1969. Breeding for high protein content and quality in barley. In: *New Approaches to Breeding for Improved Plant Protein*, Proc. Panel Rostanga, 1968, IAEA, Vienna, p. 17–21.

Jail, M.E. and Tahir, W.M., 1973. World supplies of plant proteins. In: *Protein in Human Nutrition*, (Eds.) J.W.G. Porter and B.A. Rolls. Academic Press, London, p. 35–46.

Lowry, O.H., Rosenbrough, N.J., Far, A.L. and Randall, R.J., 1951. Protein measurements with Folins reagents. *J. Biol. Chem.*, 193: 265–275.

McCarthy, T.E. and Sullivan, M.X., 1941. Rapid determination of Methionine in crude protein. *J. Biol. Chem.*, 141: 871.

Mertz, E.T., Bates, L.S. and Nelson, O.E. 1964. Mutant gene that change protein composition and increase Lysine content of Maize endosperm. *Science*, 145: 279–280.

Naismith, W.E.F. 1955. Ultracentrifugal studies on soyabean protein. *Biochem. Biophys. Acta*, 16: 203–210.

Nelson, O.E., 1969. Genetic modification of protein quality in plants. *Ad. Agron.*, 21: 171–194.

Pant, R. and Tulsiani, D.R.P., 1969. Solubility amino acid composition and biological evaluation of proteins isolated from leguminous seeds. *J. Agric. Food. Chem.*, 17: 361–366.

Prasad, P.R. and Haider, Z.A., 1985. Grow more french bean and bridge the protein gap. *Seed and Farm*, 11: 23–24.

Prasad, P.R. and Prasad, A.B. 1981a. Induced variability in methionine content in *Phaseolus vulgaris* L. In: *Prospective in Cytology and Genetics*, Proc. 3rd All India Congress of Cytology and Genetics, (Eds.) G.K. Manna and U. Sinha, 3: 679–688.

Prasad, P.R. and Prasad, A.B., 1981b. Mutagenic activity of antibiotics alone and in conjunction with alkane sulfonates. *Mutation Res.*, 84: 83–90.

Roberts, R.C. and Briggs, D.R. 1965. Isolation and characterization of the 7[s] Components of Soyabean globolins. *Cereal Chem.*, 42: 71–85.

Schons, W.J., 1970. Analysis of protein and amino acid in protein rich barley strains. In: *Improvement of Plant Protein by Nuclear Techniques*, IAEA, Vienna, p. 265–273.

Zimmermann, G. and Levy, C., 1962. Correlation between alcohol insoluble substances and Lysine availability in canned peas. *J. Agric. Food. Chem.*, 10: 51–53.

Chapter 10

Effect of Temperature on Plants with Special Reference to Wheat

Kanak Sinha[1], Bikash Patnaik[2] and Santosh Kumar[2]

[1]*Amity Institute of Biotechnology, Amity University,*
Sector-125, Super Express Way, Noida, U.P.
[2]*University Department of Botany,*
B.R.A. Bihar University, Muzaffarpur-842 001, Bihar

Introduction

Temperatures are used in connection with day length to manipulate the flowering of plants. Thermoperiod, refers to daily temperature change, Plants produce maximum growth when exposed to a day temperature which should be higher than the night temperature this allows the plants to photosynthesize and respire during an optimum daytime temperature and to curtail the rate of respiration during a cooler night-high temperature because increased respiration sometime above the rate of photosynthesis. This means that the products of photosynthesis are being used more rapidly than they are being produced. For growth to occur, photosynthesis must be greater than respiration.

The world's climatic zones are determined largely by latitude and altitude and have vegetation types established by natural selection in response to the prevailing temperature and rainfall (*Sutcliffe*, 1977). The mountains in Great Britain come treeless at a relatively low level and trees are rarely found above 600 m. However, in *New England*, U.S.A. the mean annual temperature in the lowlands is about the same as in Scotland, the tree line, *e.g.* on Mount Washington (1917 m) occurs at about 1500 m. The reason for this is that summer temperature in New England is higher than in

Scotland both in the low land and at the same altitude on mountains and the growing season in New England is longer (*Sutcliffe*, 1977). Hence the growth of trees at high altitudes on British mountains is limited by low summer temperatures.

Temperature as an Important Ecological Factor

Temperature as a Limiting Factor for Plant Distribution

Plant distribution is also controlled by winter temperature. Plants which are sensitive to chilling and frost are limited in their distribution to the minimum temperature to which they are exposed. On the other hand, many plants are unable to survive in extremely hot climates such as in the desert. For example the geographical distribution of the *stemless thistle. Cirsium acaule* in Europe from Sweden to *southwestern* England follows the isotherms for mean and maximum temperatures in July, August and September. It seems the geographical distribution of *Cirsium acaule* is determined by summer temperature (*Sutcliffe*, 1977).

Woodward (1987), has stated that the absolute minimum temperature is crucial in explaining the distribution of plants.

The likely explanation for the high-temperature distribution limits for the species for cold climates is competitive exclusion by species from warmer climates (Woodward, 1987). For example *Sedum rosea*, a species from high altitude were grown in competition with *Sedum telephium*, a species restricted to low altitude. After a period of three years it was observed that S. *telephium* became extinct at the highest altitude, while S. *rosea* was excluded from the lowest altitude (Woodward, 1987).

In a climatic region those plant species are successful which have evolved mechanisms to resist winter cold or summer drought (Cooper, 1963). The study of climatic adaptation involves the study of the pattern of variation in existing climatic races and relating this pattern to the selective action of local climate. Winter is the most favourable growing season in the Mediterranean environment. However, summer drought limits plant growth. Therefore developmental seasons have been selected which allow active growth in winter and the survival for the summer drought in the form of seeds, as in the winter annuals *Lolium rigidium* and *Trifolium subterranean*, or through summer dormancy as in *Phalaris tuberosus* or in Mediterranean populations of *Dactylis glomerata* (Knight, 1960) and *Lolium perenne* (Salisbury, 1961).

The geographical distribution of some plants may be determined by diurnal fluctuation of temperature during the growing season (Sutcliffe. 1977) *Zinnia* for example require warm temperatures over the range of 20-30°C and since such conditions exist in Pasdena, California, U.S.A. from July to September, the plants grow well during these months (Sutcliffe, 1977). Chinese Asters on the other hand require lower day and night temperatures than *Zinnia* and these conditions occur in spring and autumn (Sutcliffe, 1977).

The timing of flowering and seed production shows close adaptation to the local climate. In the Mediterranean climate flowering usually takes place resulting in seed production at the beginning of the dry season soon after the water supply becomes exhausted. In the Mediterranean annuals'. R. *rigidium* (Coopet, 1959, 1960) there is little or virtually no winter requirement for floral induction, and initiation can occur in the comparatively short photoperiods of the Mediterranean winter followed by flowering in early spring. Even within a Mediterranean environment, however, the possible growing season varies, due to the duration and amount of winter rainfall. Donald (1960) cited by Copper (1963), has suggested that the distribution of *T. subterranean* in Australia is limited by three climatic boundaries; aridity, heat and cold. The arid boundary is determined by the length of the effective rainfall season, and its position thus varies with the life cycle of the variety: the early-flowering Dwalganup, for instance, requires a growing seasons of about 5 months while the midseason Mount Barker needs seven months. The warm boundary is set by the low temperature requirement for floral induction, which again varies with the variety. Talbrook requiring temperatures below 12°C, while Dwalganup will flower up to 24°C. The cold boundary is found only at high altitudes in southeast Australia, where frost may damage vegetative growth or flowering.

In northern Europe Summer drought is not usually limiting, and the long days of summer provide the optimum conditions for seed production. In north temperate varieties of such forage species as *D. glomerata* and *L. perenne* floral initiation and elongation usually require long photoperiod (Gardner and Loomis, 1953).

Seed dormancy provides a further mechanism for adaptation to the climatic range. Cultivated forage species are sown in the season most favourable for them and therefore dormancy in disadvantageous.

Effect of Temperature on the Physiological Processes of Plants

The understanding of the physiological basis of climatic adaptation involves the analysis of local climate into components such as light energy and photoperiod day and night temperature, and the availability of moisture and a study of the effects of these components on individual plant processes such as the rate of photosynthesis and respiration, leaf and bud initiation and expansion, and the induction, initiation, and expansion of the inflorescence (Cooper, 1963).

Temperature influences metabolism and growth of plants. The optimal temperature for growth differs from species to species. For example green algae found in arctic and antarctic waters survive indefinitely and grow at temperatures which barely exceed 0°C. On the other hand, blue-green algae found in hot-springs can grow at temperatures as high as 85°C (Sutcliffe, 1977). The optimum temperature for growth not only varies between different species but also varies between different organs of the same plant. Changes in the temperature of one part of the plant, for example the roots, may affect the growth of another (*e.g.* the shoot) even when the temperature of the latter is kept constant (Sutcliffe, 1977).

Climatic races are usually closely related to their climatic origin from the observation of pattern of vegetative growth under controlled light and temperature regimes (Ortiz *et al.*, 2008). The effect of day and night temperature on the growth of contrasting species and hybrids of *Poa showed* a general correlation with climatic origin (Hiesey, 1953). *Poa pratensis* from middle latitudes. For instance, produced most bulk at cool night and day temperatures, but the arctic form from Lapland grew better at fairly high day and night temperatures, possibly corresponding to the small diurnal temperature range of the arctic summer. The Mediterranean *Poa scabrella,* on the other hand, grew well on cool days and cool nights, and became dormant at high temperature.

The variation in rates of photosynthesis and respiration could lead to variation in vegetative growth in response to temperature. It is however important to distinguish between plants growing without competition for light and plants growing in dense stands in which self shading is usually operating (Watson, 1958).

Laude (1953), who studied summer dormancy in forage species found that several perennial species such as *Phalaris tuberosa* and *Oryzopis miliacea* which became summer dormant in California due to water stress, would continue to grow in irrigated, while others such as *Poa scabrella, Poa secunda,* required lower temperatures and shorter photoperiods to grow and would not respond to irrigation, Knight (1960), found that dormancy during summer occurs in many Mediterranean populations of *D. glomerata.* However, here some species respond and start to grow after irrigation while others remain dormant even after supplying water. It is suggested that such summer dormancy may occur due to occasional but unreliable summer rains. Most temperate perennial forgae species require verbalization treatment *i.e.* exposure to winter conditions of cold and/or short days floral induction, while Mediterranean species have little requirement for this. Depending on the genus, the pattern of inductive response varies. In *Lolium* for example induction can be brought about either by cold (0-5°C) or by short days and seedlings are capable of responding to both these factors soon after germination (Cooper, 1960).

Plant Productivity Affected by Temperature

Temperature influences nearly every aspect of the biology of plants (Nobel, 1988). Temperature and water are major determinants of plant productivity (Grace, 1988). Tomatoes, for example, grow faster when the day time temperature is 26°C and the night time temperature is 17°C. The growth of tomatoes is reduced at any intermediate temperature or at a constant temperature of 26°C (Sutcliffe, 1977).

Root dry weight, leaf area and leaf concentrations of most nutrient elements were all reduced as the root temperature was raised to 35°C and to 38°C in particular in cucumber plants (Du–Yc and Tachibana, 1940).

Plant productivity can be affected to a large extent by variations in temperature such as those that exist between the north and south of Britain, or even between north and south aspects of a hill (You *et al.,* 2009). During the growing season, in a northern temperate climate, a 1°C rise in temperature is expected to increase plant productivity by about 10 per cent, provided that other factors such as water and nutrients do not become limiting (Grace, 1988). Cold places contain fewer species compared with warm places and have biological life forms which are dwarf and standing crop and productivity are very low (Box, 1981 cited by Grace, 1988). Monteith (1981 a cited by Grace, 1988), found a negative correlation between the yield of wheat crops in creation parts of southern England and the mean temperature of the period May, June and July. He suggested that temperature often determines the rate at which the crop develops from sowing, through phonological stages to flowering. Thus, in warm years, the duration of the crop is less and as a result totals photosynthesis and dry matter production is reduced. It is also possible that warm years are usually dry years and crop growth in southern England is limited by water stress.

A change in altitude also brings about change in weather variables, such as temperature. Prince (1976) grew barely at high altitude in the English Pennies and observed that it produced seeds one month later than at low land sites, the grains being reduced in weight. The effect of temperature on tree productivity is obtained by the analysis of radial increments taken from the core of the stem. Productivity is in this case radial trunk growth and not yields. Over northern Europe, the growth of Scots pine *Pinus sylvestris* has been shown to be strongly correlated with summer temperature (Hughes *et al.*, 1984).

Watts (1971), demonstrated that the temperature of basal meristems was more important in determining extension growth of grasses. The rate of photosynthesis is less sensitive to temperature than cell division and extension and that growth is less dependent on photosynthesis. It is possible that temperature of meristems is more closely affected by fluctuations in radiation and wind speed than they are by air temperature. Linacre (1964, 1967 cited by Grace, 1988), observed that broad-leaved herbaceous vegetation is not particularly well coupled to air temperature, especially in bright sunshine and very cold or very hot environments. In many cases the leaf temperature of plants growing in hot climates is a few degrees cooler than the air temperature as a result of transpiration (Smith, 1978).

Reamer (1935) cited by Grace (1988), was one of the first authors to investigate the quantitative relationship between temperature and plant performance. Soon after inventing his thermometer, he used it to investigate the effect of temperature on the dates of harvest of grapes and wheat in France. He proposed the 'heat unit' law, that a fixed 'quantity of heat' is needed for a plant to reach a particular stage of development. This 'law', though frequently violated, has proved useful in many instances, leading to the widespread use of degree-days as an index of the extent of the growing season (Nuttonson, 1955 cited by Grace, 1988). Excellent linear relationships have been observed by crop physiologists between state of development and accumulated temperature in a wide range of species including north temperate trees (Landenberg, 1974), temperate crops (Russell *et al.*, 1982) and tropical graminaceous crops (Ong, 1983 a, b; Hadley *et al.*, 1983; Roberts, Hadley and Summerfield, 1985).

Thorn *et al.* (1967), observed that the rate of photosynthesis is much less sensitive to temperature than is the rate of leaf extension growth. In many plants for instance a reduction in temperature with the onset of winter causes an increase in the pool size of soluble carbohydrate (Eagles, 1967; Sutcliffe, 1977). There is evidence to suggest that the effect of temperature on growth is greater than the effect of temperature on photosynthesis. Warren-Wilson (1966), showed that the specific leaf area is the variable which is most sensitive to temperature with the ratio of area to weight increasing with temperature.

Effect of Temperature on Wheat *(Triticum aestivum L)*.

Wheat is grown throughout the world, from the borders of arctic to near the equator, although the crop is most successful between the latitudes of 30° and 60°

North and between 27° and 40° South. In altitude, it ranges from sea level to 3050 m (10,000 ft.) in Kenya and 4572 m (15,000 ft.) in Tibet. Cultivated varieties, which are of widely different pedigree and are grown under varied conditions of soil and climate, show wide variations in characteristics. Wheat grows best in subtropical, warm temperate and cool temperate climates. An annual rainfall of 229-762 mm, falling more in spring than in summer, is suitable to its growth. The mean summer temperatures should be 56°F (13°C) or more (Kent, 1975).

The two major constraints to wheat production are the occurrence of late spring frosts and the onset of the dry season and drought (Miglietta and Porter, 1992). For winter wheat the emergence to floral initiation phase is measured in thermal time modified by photoperiod and vernalisation factors that reflect their faster development in long days and requirement for a period of low temperatures, winter wheat (*Triticum aestivum L.*) cultivars produced in the semiarid environment of the Canadian priarises are subjected to variable water stresses (Entz and Fowler, 1990). As stress increased, the grain yield for Norwin declined at a significantly greater rate than that of Norstar. Several researchers have also reported that compared with tall types, semidwarf wheat cultivars are more productive under favourable conditions (Allan, 1986) but more susceptible to drought and high temperature stress (Liang and Fisher, 1977). The studies of the above researchers show that the yield of the cereals plants can be affected by drought, and that cultivar variation in drought tolerance for these traits occur.

Spring wheat (*Triticum aestivum L.*) is a major crop of the northern great plains (U.S.A.). Hexaploid wheats are of temperate origin (Harlan and Zohar, 1966). High temperature stress during the grain-filled period is a major constraint to increased productivity of spring wheat (Shanahan *et al.*, 1990). Like most species of temperate origin, wheat does not tolerate prolonged exposures to temperatures exceeding 35°C (Gust and Chen, 1987 cited by Shanahan *et al.*, 1990). Heat stress during the grain fill hastens maturation of the crop, resulting in small shrunken grains.

In parts of the world which have mild winters and hot summers, wheat is generally sown in the autumn when temperatures are falling, and harvested in early summer prior to the hottest months of the year. In India for example, sowing is done in October to November wile the grain is harvested in March to April. Late sowing reduces the yield because grain filling occurs during the heat of summer, while early sowing curtails vegetative development thereby reducing the crop's yield potential (Owen, 1971 cited by Bagga and Rawson, 1977). The use of short duration cultivars could used to avoid high temperatures at the beginning and end of the season, although the approach ensures more reliable year to year yields, mean yields are relatively low, as short duration cultivars fail to benefit from the occasional longer season. As alternative approach, which would theoretically result in more stable and higher yields, would be to use high-temperature tolerant cultivars with a longer duration. However, little is known of what plant characters are associated with high temperature tolerance, whether these characters are common in wheat cultivars or whether tolerance in a cultivar at one stage of development implies tolerance at another. A study was attempted by Bagga and Rawson (1977), to determine if and why there are differences among the three cultivars of wheat *Kalyansona, Condor* and *Janak* in their responses to temperature. *Kalyansona* was relatively unresponsive to temperature during the floret phase, being little affected in the sizes of upper leaves, in floret production and grain set, in overall plant growth or in grain yield. The sole character to respond to temperature in this cultivar was kernel weight, which declined with increasing grain phase temperature. In contrast, *Condor* demonstrated marked plasticity during the floret phase in all plant characters measured. Its plasticity was such that, at the lower temperatures, it outyielded *Kalyansona* by a substantial margin while at the higher temperatures its yield was relatively poor. On a plant basis, *Janak* performed similarly to Condor. Rates of photosynthesis were relatively unaffected by temperature in any cultivar.

This wide range of response among three superficially similar cultivars has promising implications for the tailoring of cultivars for different Zones.

High temperature stress during the grain-fill period is a major constraint to increased productivity of winter wheat grown in the central and southern Great Plains of the U.S.A. and in other areas of the world (Wardlaw *et al.*, 1989 cited by Saadalla, 1990). High temperature (up to 30/25°C day/night) reduced grain yield of wheat (Saadalla, 1990). Wiegand and Cuelar (1981), reported significant reduction in grain yield when temperature averages during grain fill were above 15°C. It has been demonstrated that high temperature retards conversion of sucrose to starch in developing grains of wheat (Bhullar and Jenner, 1985, 1986 and Rijven, 1986).

Photosynthetic activity decreases rapidly when temperate species are exposed to heat stress during reproductive development. Photosynthetic rates of leaf discs decreased faster at 35°C than at 20° (Hording *et al.*, 1990). They concluded that high temperature initially accelerated thylakoid component breakdown, an effect similar to normal sensescence patterns. High temperature also reduces grain growth in wheat and reduces productivity (Harding *et al.*, 1990).

Photosynthetic rates and thylakoid activities are adversely affected by high temperatures in wheat (*Triticum aestivum* L.). Ten genotypes from major world wheat-producing regions were grown under moderate (22/17°C day/night) and high (32/27°C day/night) temperatures for two weeks as seedlings. High temperature decreased mean photosynthetic rates and mean total biomass (Al-Khatib and Paulsen, 1990). Photosynthesis rates of wheat (*Triticum aestivum* L.cv. Len) seedlings declined gradually after temperature increased from 22 to 42°C (Al-Khatib and Paulsen, 1990).

Heat Effects on Protein Biosynthesis

Plant tissue generally responds to sudden increases in temperature by curtailing or abolishing normal protein synthesis and producing new polypeptides known as heat shock proteins (HSP) (Ougham and Howarth, 1988). In almost all eukaryotic tissues, the synthesis of heat shock proteins in induced rapidly (within minutes) when the temperature is raised above the requisite threshold. This threshold varies from species to species; for example, in maize it is approximately 35°C (Cooper and Ho, 1983; Cooper *et.al.*, 1984 and Cooper and Ho, 1987). Above the threshold, the exact nature of the response depends on both temperature and duration of the heat treatment (Kalra *et al.*, 2008). For example, in millet seedlings grown at 35°C, heat shock proteins are first detectable at about 40°C. At 45° HSP synthesis is maximal and normal protein synthesis is greatly reduced. At 50°C, there is a small amount of residual HSP synthesis but normal proteins are no longer synthesized. At 55°C protein synthesis is abolished completely. For temperate plants lower temperatures are sufficient both to induce maximal HSP synthesis and to abolish protein synthesis (Ougham and Howarh, 1988).

This so-called heat shock response has been most extensively studied in the fruitfly, *Drosophila melanogaster*, the organism in which this phenomenon was first observed. When *Drosophila* cells are shifted from their normal growth temperature (25°C) to an increased temperature (37°C), there is a cessation of normal protein synthesis with the concomitant synthesis of a small set of heat shock proteins. The message coding for the heat shock proteins result from new transcription and are preferentially translated during heat shock. That is normal mRNA are rapidly cleared from the ribosomes to make way for the translation of heat shock mRNA (Cooper and Ho, 1983).

While high temperature activates transcription of heat shock genes, it completely represses the transcription of nearly all the other genes. When Drosophila cells are returned to 25°C, normal transcription and translation resume, and heat shock protein synthesis gradually declines.

An important aspect of heat shock protein synthesis in a tissue is its effect upon thermotolerance. For example, if soybean seedlings grown at 30°C are exposed to 45°C for 2h. little protein synthesis of any kind takes place and the seedlings do not survive upon transferring them to 30°C. However, if seedlings are exposed to 40°C for 2h during which heat shock protein synthesis occurs and then transferred to 45°C, seedlings produce large amount of heat shock proteins, thus acquiring thermotolerance and upon returning the seedlings to 30°C normal growth resumes (Key *et al.*, 1985b as cited by Ougham and Howarh, 1988). A similar effect has been observed in sorghum (Oughum and Stoddart, 1986.) Lin *et al.* (1984), demonstrated the development of thermal tolerance in "Wayne" soybean seedlings. Hence, it was inferred that induction of heat shock protein synthesis leads to thermotolerance in tissues (Donovan *et al.*, 2005).

Krishnan *et al.* (1989), reported that plants respond to high temperature stress by the synthesis of heat shock proteins and thus acquiring thermotolerance to lethal temperatures. Chen *et al.* (1982), also observed genetic diversity in the thermal tolerance of beans and tomatoes.

Effect of Temperature on Cell Division

Cell growth is sensitive to temperature (Sutcliffe, 1977). Temperature affects both mitosis and cytokinesis. It has been observed that in *Vicia faba* roots both mitosis and cytokinesis proceed very slowly at temperatures below 3°C and at a faster rate at about 25°C (Sutcliffe, 1977). The cell cycle was classified into its component phases for the first time by Howard and Pelc (1953). They divided the interphase period into Pre-DNA synthetic phase (G1), DNA synthesis (S-Phase) and post DNA synthetic phase (G2). A G0 phase has been proposed to classify non-cycling cells with the 2C amount of DNA. When there is an increase in temperature the rates of cell division increases. However, when the temperature is increased beyond a certain limit disruption of cell division occurs. Barlow and Adam (1989), found that the rate of elongation of primary root of *Zea mays* grown in solution culture at 5°C was about only one percent of the rate grown at 20°C, the apical meristem was also short and there was also an increase in the rate of elongation of the primary root, and an increase in meristem length probably because of the recovery of mitosis *i.e.* cell division taking place at a faster rate in the quiescent centre. However, the recovery depends on how long the roots were kept at 5°C, the longer the exposure to 5°C, the greater is the time taken to restore the balance between cell division and cell differentiation. This is because cell differentiation was being less affected than cell division and the apical differentiation was being less affected than cell division and the apical meristem losing cells at a faster rate than that were being produced. The amount of cell division in the quiescent centre is related to the duration of the cold treatment.

Evidence suggests that rates of cell division alter with tissue and species specific manner (Francis and Barlow, 1988). For example, Taylor and Clowes (1978) showed that in *Zea mays* G1 is eliminated in short cell cycles induced by higher temperature but in *Allium sativum* there is always a G1 even if it is less than one hour in duration. In the latter species the optimal temperature for mitosis was slightly higher than that for cdt. The highest rate of cell production in the meristem proper were at 24-30°C, S Phase and M increasing the proportion of the cell cycle they occupy as temperature increases but G1 and G2 shortening. In the cortical regions the proportions of the cell cycle occupied by the different phases were different. Webster and Macleod (1980) showed that cell cycles in the cortical and boundary regions of the meristem may be slower than in the meristem proper, as cell begin to differentiate.

The nuclear cycle provided for the replication of chromosomes, new copies of which are distributed to the daughter cells arising from cytokinesis. The most important variables to which the cycle is susceptible are temperature, oxygen availability and moisture level (Barlow, 1987).

Nuclei can reproduce and divide over a wide range of temperature and the duration of the nuclear cycle decrease with increasing temperature (Lopez-Saez *et al.*, 1966). The lower temperature limit for the cycle *e.g.* in barley is about 0.5°C (Grif and Valovich, 1973) and the upper limit is a little above 35°C. The optimal temperature for the nuclear cycle varies from species to species. In onion roots the cycle is fastest at 27°C and remains constant above this temperature; the nuclei can still divide at 35°C but division ceases at 40°C. Maize shows an inverse relationship between division rate and temperature up to 35°C.

Temperature related changes in division cycle are accompanied by proportional in the duration of the component phases G1, S, G2 and mitosis (Gonzalez-Fernadez *et al.*, 1971).

At the molecular level, events of the S phase are the best characterized. During S discrete lengths of the DNA double helix (replicons) are replicated in a particular sequence. Temperature influences the rate of movement of the DNA replication fork along the replicon (Van'tHof, Bjerknes and Clinton, 1978). In *Helianthus annuus* the average fork rate is 6 μm/h at 10°C, 8 μm/h at 20°C and 11.5 μm/h at 35°C. Below 15°C the S phase becomes protracted because the initiation of replication is less efficient.

Francis and Barlow (1988) plotted the percentage of the cell cycle occupied by G1, S, G2, and M in root meristems of five species exposed to various temperatures under controlled conditions. In all five species the total duration of the cell cycle decreased with increasing temperature but the phase did not increase proportionally. For example, in *Pisum sativum, Helianthus annuus* and *Triticum aestivum* the proportion of the cell cycle occupied by G1 progressively decreased with increasing temperature. However, in *Allium cepa* the proportions of the phases remained constant. They hypothesize that the 2C DNA amount confers maximum protection to the genome against the stress of low temperature. This may be because temperatures below 5°C may irreversibly damage cell division associated machinery, but a cell resting in G1 would remain unaffected and be able to resume normal division upon the return of more favourable temperatures. Barlow and Rathfielder (1985) showed that in the primary root meristems of *Zea mays* nuclei synthesize DNA and reach mitosis at low temperature but arrest in late G2 because they can progress no further due to pertubation of spindle formation. Wimber (1966) showed that in *Tradescantia Paludosa* G1 was shorter at 30°C than at 21°C and further reduction to 13°C resulted in trebling of the durations of G2 and M and doubling of S-Phase duration compared to 21°C.

Cribber *et al.* (1993), working on natural population of *Dactylis glomerata* of contrasting latitudina origins found that the duration of the cell cycle and the duration of the cell doubling time from high latitude/altitude was relatively less sensitive to a shift to the lower temperature than that in the lowest latitude/altitude.

There are numerous reports of toxic metal reducing the number of mitotic cells in root meristems (Powell, Davies and Fancis, 1986a). A 4-day exposure to Zn resulted in marked cellular changes in the root apices of *Festuca rubera* (Powell *et al.*, 1986a, 1986b and 1988). Exposure to Zn reduced root length to a great extent in a Zn-sensitive cultivar (S59) than in a tolerant cultivar (Merlin) (Powel *et al.*, 1980a). Accompanying the Zn-induced reduction in root length was a decrease in length of the apical meristem. Root hair formation and xylogenesis occurred closer to the root tip in Zn-treated root of the Zn sensitive cultivar (Powell *et al.*, 1988) Mitotic index was reduced and cell doubling time increased with increasing Zn concentration, these effects were more apparent in the sensitive cultivar, S59.

Root Growth and Development

The growth of root involves the process of cell proliferation, cell elongation and cell differentiation. In order to investigate the effect of temperatures on root growth it is therefore necessary to understand the processes involved in the normal functioning root and low temperatures affect each of these processes.

The root systems despite displaying varying forms their continued elongation depends on the division and subsequent extension of cells in the apical meristem (Russell, 1977).

The root apical meristem is a complex and integrated tissue, mitotic cell division of meristematic cell gives rise to the organised tissues which ultimately make up the mature functional root. A meristem where the root cap is distinct from the rest of the root, its boundary wall being well defined by relatively thick cell walls; the epidermis of the root is formed by the outermost layer of the cortex, is termed a closed meristem *e.g., Zea mays.* In an open meristem, the boundary between the root cap and the rest of the root is indistinct and the innermost layer of the root cap forms the root epidermis *e.g. Helianthus annuus* (Clowes, 1981). Elongating roots are usually regarded as possessing four regions; the root cap, the meristematic region, the region of cell elongation and the region of differentiation and maturation. The meristematic region typically consists of numerous small, compactly arranged, thin walled cells, almost completely filled with cytoplasm.

The meristem consists of two major proliferating populations: first, the proximal meristem which generates cells that go on to differentiate the mature tissue and second, a smaller population that generates cells that form the root cap. Between the two is the quescent centre: a population of cells that divide at a much slower rate compared with the surrounding cells (Clowes, 1954, 1956, 1971).

The founder cells in the QC do divide, albeit at a slow rate, and must therefore be displaced from the QC and assume a role as an initial cell giving rise to the cell population of the root. During steady-state root growth the QC remains constant in size and cell number. In the root meristem of *Allium cepa.* the number of divisions left to a cell decreased as it moved further away from the QC (Gonzalez-Fernandez *et al.*, 1968). In an actively growing root, the cells in the proximal or distal meristems have determinate reproductive life span as cells are gradually displaced by the products of cell division, whilst the derivative cell within the QC have an indeterminate life span (Barlow, 1976).

The QC is believed to be organiser of the cell patterns within a root apex. Besides their role in maintaining cell division in the root meristem, the founder cells, together with the cells immediately surrounding them, may constitute a pattern of cells that is self perpetuating. Founder cells may constitute a pool of cells whose proliferative potential is not impaired by the passage of time (Barlow, 1976). Decapitated apices of *Allium cepa* from which the QC was removed failed to regenerate a new apex (Gonzalez-Fernandez *et al.*, 1968). Reinhard (cited in Barlow, 1976), demonstrated that a new root tip of *Pisum sativum* could only be regenerated from the meristematic cells which included the QC. When Clowes and Hall (1966), irradiated roots of *Vicia faba* with gamma-irradiation in the region of the QC, so that the QC cells cycled at the same rate as the cells in the rest of the meristem, they found that considerable disorganisation of the lineages occurred.

Impairment of the division cycle of initial and derivative cells frequently cause the cells in the QC to speed up their mitotic cycles (Clowes, 1967). Consequently it is possible for the founder cells to repopulate the meristem if other cells are damaged.

The root meristem is characterized as a steady-state system maintaining its growth pattern over a period of time. Important properties in such a system are constant meristim size, cell number, rates of cell production within the meristem and cell loss from the meristem and asynchronous cell division. In an asynchronously proliferating population of cells, such as the root meristem, cell number increases exponentially. However, the rate of cell production in the meristem is balanced by the rate of transit cells to the non-meristematic regions of the root at the proximal margin and to the root-cap at the distal margin (Webster and Macleod, 1980). Therefore cell number tends to remain constant with exponential growth occurring only for any instant in time within the proliferative population.

Age distribution in a proliferating cell population is not strictly related to the relative duration of phases of the cell cycle *i.e.* the proportion of cells in any stage in the cell cycle is not equal to the relative duration of that stage. Each cell division gives rise to two daughter cells and thus twice as many cells at the start of a cycle as there are at the end. Only at the meristem margins in every cell division balanced by cell loss, thus within the proliferating population the relative frequency of cells in particular stages of the cell cycle declines experientially with time across one cell cycle (Natchtwey and Cameron, 1968).

Lateral Root Formation

Lateral root primordial arise indigenously. They initiated in the pericycle of the parent root (Lyndon, 1979). During initiation of a leateral root a group of pericycle cells undergoes periclinal and anticlinal divisions ultimately forming the lateral root primodium. The complexity of factors determining lateral root formation has been shown by Torry (1959). The absence of lateral roots close to the apex suggests that the apical meristem produces substances which inhibit the formation of lateral roots. The time between initiation of a primordium and subsequent emergence can range from about 4-8 days (Blakely, Rodaway and Hollen cited in Webster and MacLeod, 1980 and Hackett and Stewart, 1969).

Root Growth and Temperature

Barlow (1987), reported that root systems, often with very contrasting architecture, thrive in a diversity of environments. Stress may cause cessation of growth in the root, and the resumption of growth is often marked by stimulation of division of cells in the quiescent centre. For example, Barlow and Rathfielder (1985), found that in *Zea mays* roots, exposure to low temperature caused inactivation of the root cap initials which were then replaced by division of neighboring cells in the quescent centre. Also, isolated cells in the proximal meristem which were killed by the low temperature, caused re-orientation of patterns of cell division, so that the dead cells were crushed and replaced. Thus the quiescent centre has a capacity to withstand stress and remain undamaged, so that upon activation it can regenerate as apex similar in construction and pattern of differentiation to the damaged part (Barlow, 1987).

Roots are subject to variation in temperature, chemical environment and soil compaction and yet they are broadly homoeostatic and are generally able to accommodate changes without alteration to the basic structure (Barlow, 1987). However, the changes in the physical environment can bring about changes to roots, for example in meristem length or cellular length or in the architecture of the whole system. One of the most obvious differences is in rate of extension growth and overall length achieved. Lopez-Sacz *et al.* (1969), reported that the growth rate of *Allium cepa* roots increased as temperature was increased from 5-35°C and then decreased at 35°C. Pahlavanian and Silk (1988), reported that cell length increases with temperature at all locations. However, the relative cell length change in the 3 to 5 mm region in invariant with temperature.

Korer and Renhardt (1987) have shown that high altitude species have a greater proportion of their biomass allocated to roots due to reduced stem height and increase in fine roots. Berntson and Woodward (1992) also found that in Senecio vulgaris, high CO_2 levels and increased water supply, as would be found at high altitude caused the roots become more branched, longer and more horizontal. Thus, variation in the physical environment of the roots influenced the volume of soil exploited by the plant.

Wheat was grown by Aston (1987) in temperature controlled water baths in order to regulate the temperature of the roots, the growing apex and the zone of leaf extension. Twenty-two alternating day and night temperatures varying between 26 and 2°C were studied. Dry matter production was found to be exponentially related to the time corrected mean daily temperature of the apex, leaf extension zone and roots. These results confirm the suggestion that the improved early vigour of wheat sown by conventional cultivation practices with a minimum of surface residues compared to direct drilled wheat, could at least partially have been due to the different patterns in soil temperature.

Hay and Tunnicliffe Wilson (1982), suggested that soil temperature was a major factor determining early growth. Aston and Fischer (1986) measured soil temperatures at various depth under wheat crops sown by different methods of cultivation. They also suggested that the soil temperature regime from the soil surface down to a depth of 1 cm was major factor determining early wheat growth.

Temperature exerted a strong influence on the partitioning of new growth to the roots and the shoots. In experiments where root temperature was varied while shoot temperature was held constant and near optimal, there was a distinct "optimum root temperature" where shoot growth was maximal and the root/shoot ratio was minimal. For species of temperate origin the optimum was between 20 and 30°C. Growth was reduced at low temperature, but the growth occurred at near freezing temperature. Subtropical species typically have a higher optimum root temperature and growth is often totally inhibited at root temperatures well above freezing (Lange *et al.*, 1981).

Marilyn and Intyre (1990), examined root growth, development and frost resistance in winter rye (*Secal cereale* L. cv. *Puma*). Rye roots grown 5°C develop first order lateral roots, differentiate metaxylem vessels and suberize endodermal cell walls more slowly than roots grown at 20°C.

Engels and Marschner (1990) grew maize seedlings for 10 to 20 days in either nutrient solution or in soils. Air temperature was kept uniform for all treatments, while root zone temperature (RZT) was varied between 12 and 24°C. Shoot and root growth were decreased by low RZT to a similar extent irrespective of the growth medium.

Rufty *et al.* (1981) found that exposure of healthy, vegetative soybean plants to different root temperatures resulted in alterations of roots. They found that greatest root growth occurred at root temperatures of 24°C. When net CO_2 assimilation of leaves was decreased by low light intensity roots became more sensitive to increasing root temperature, decreasing in size at 24°C and 30°C relative to roots at 18°C.

The Greenhouse Effect on Plant Growth

The more recent increase in greenhouse gases since pre-industrial times can be related to human activities. Climate models predict a significant global warming of several degrees within the next century if the industrial emissions increase unabated (Roecker, 1992). The major matter of scientific and public concern is the additional greenhouse effect caused by human activities. The gradual increase of atmospheric greenhouse gases will create an imbalance between the solar radiation absorbed by the earth-atmospheric system which remains nearly unchanged and the reduced energy loss by increased atmospheric trapping of infra-red radiation. The excess energy will heat the climate system in order to restore the energy balance.

However, Goudriaan (1992), states that a rising concentration of atmospheric CO_2 will stimulate plant growth and productivity through increased photosynthesis and the improved efficiency of utilisation of scares resources. Most notable the efficiency of plant water use in increased, so that the growth rate can be stimulated without an increase in the water demanded for transpiration.

Higher temperatures, larger water deficits and high light stress are likely to occur in conjunction with elevated atmospheric CO_2. Chaves and Pereira (1992), stress that doubling CO_2 concentration in the air may improve carbon assimilation and compensate partially for the negative effects of water stress even if it is assumed what a down-regulation of the photosynthetic process occur as a result of acclamation to elevated CO_2. Sinclair (1992), suggests that a positive response to increased CO_2 concentration, for example requires an increase in plant uptake of the total amount of minerals. Consequently it is very difficult to predict the plant growth response to climate change because of the large uncertainty about mineral availability. Carter *et al.* (1992), suggested that a mean annual temperature increase of only 1°C would open up large areas in southern England, the low countries, eastern Denmark, northern Germany, and northern Poland to potential grain maize cultivation. An increase of 4°C would move the limit into central Fennoscandia and northern Russia.

Woodward and Friend (1988), state that a thorough understanding of the mechanisms which control the temperature responses of growth is central to studies of plant distribution and plant breeding for increased productivity (Rosell and Coller, 2009). The understanding of normal responses of plant growth processes to temperature is necessary to an understanding of the ecophysiological processes of adaptation. Such knowledge would help in improving the accuracy of predictions of the effects of environmental change, both in agricultural and wild species, the latter a possible genetic resource which could be tapped if better understood. Sinclair (1992) suggests that a positive response to increased CO_2 concentration. For example requires an increase in plant uptake of the total amount of minerals.

Importance of Studies on Wheat

Origin of Bread Wheat

Triticum aestivum L. belongs to the family Poaceae and is a monocotyledon. The chromosome number (2n) is 42 at a hexaploid level (6x) and has a DNAC value of 31.6 pg. (Bonnet and Smith, 1976).

Bread wheat, *Triticum aestivum* (2n=6x=42, AABBDD), derives from a hexaploid which seems to have originated from domesticated emmer (*Triticum dicoccum.* 2n=4x,AABB) and a wild grass (*Aegilops suquarrosa* auct. non. L., 2n=2x, DD) (Zeven, 1979). Zeven (1979), came to the conclusion that bread wheat was spread to N.W. Europe as an admixture of emmer. The mixture must have reached the Balkans in ca. 6000 BC and N.W. Europe ca. 4000 BC. Archeological finds show the presence of a few bread wheat grains in emmer grains.

Vernalization and Photoperiodic Requirement of Wheat

The two decisive traits which govern development rates of varieties of common wheat are vernalization requirement and photoperiodism (Martinic, 1973). Sometimes a good variety may give poor yield when introduced into a new region because its vernalization requirement is not met or the day length is too short for its normal development. Spring wheat varieties do not respond at all to low temperature. While those that are classified as winter wheat types require vernalization treatment *i.e.* exposure to low temperature treatment for a certain period of time which varies from 20-30 to more than 60 days, depending upon variety before sowing in late spring (Martinic, 1973). It has been observed that in the unvernalized varieties when sowing is done in late spring, heading is delayed for a certain number of days and sometimes heading fails to occur (Martinic, 1973). Spring wheat seeds are usually sown in spring and winter wheat seeds are usually sown in late autumn (Kent, 1975).

In countries which experience severe winter such as the Canadian prairies and the steppes of Russia wheat is generally sown in early spring and harvested before the first frosts of autumn. Winter wheat is usually sown in places where excessive freezing of a soil does not occur *e.g.* north-western Europe. The grain germinates in the autumn and grows slowly until the spring while spring wheat in those countries maximum rainfall is in spring and early summer, and maximum temperature in mid- and late summer (Kent, 1975).

References

Al-Khatib, K. and Paulsen, G.M., 1989. Enhancement of thermal injury to photosynthesis in wheat plants and thylakoids by high light intensity. *Plant Physiology*, 90: 1041–1048.

Al-Khatib, K. and Paulsen, G.M., 1990. Photosynthesis and productivity during high-temperature stress of wheat genotypes from major world regions. *Crop Science*, 30: 1127–1132.

Allan, R.E., 1986. Agronomic comparisons among wheat lines nearly isogenic for three reduced-height genes. *Crop Science*, 26: 707–710.

Aston, A.R., 1987. Apex and root temperature and the early growth of wheat, *Australian Journal of Agricultural Research*, 38: 231–238.

Aston, A.R. and Fischer, R.A., 1986. The effect of conventional cultivation, direct drilling and crop residues on soil temperatures during the early growth of wheat at Murrumbateman, New South Wales. *Australian Journal of Agricultural Research*, 24: 389–60.

Bagga, A.K. and Rawson, H.M., 1977. Contrasting responses of morphologically similar wheat cultivars to temperatures appropriate to warm temperature climates with hot summers. *Australian Journal of Plant Physiology*, 4: 877–887.

Barlow, P.W., 1976. Towards an understanding of the behaviour of root meristems. *Journol of Theoretical Biology*, 57: 433–451.

Barlow, P.W., 1987. The cellular organization of roots and its response to the physical environment. In: *Root Development and Function*, (Eds.) P.J. Gregory, J.V. Lake and D.A. Rose Society for Experimental Botany, 40: 1–27.

Barlow, P.W. and Adam, J.S., 1989. The response of the primary root meristem of *Zea mays* L. to various periods of cold. *Journal of Experimental Botany*, 40(210): 81–88.

Barlow, P.W. and Rathfielder, E.L., 1985. Cell division ad regeneration in primary root meristems of *Zea mays* recovering from cold treatment. *Environmental and Experimental Botany*, 25: 303–314.

Bennett, M.D. and Smith, J.B., 1976. Nuclear NDA amounts in angiosperms. *Philosophical Transactions of the Royal Society*, 274: 227–274.

Berntson, G.M. and Woodward, F.I., 1992. The root system architecture and development of *Senecio vulgaris* in elevated CO_2 and drought. *Functional Ecology*, 6: 324–333.

Bhullar, S.S. and Jenner, C.F., 1985. Differential response to high temperature of starch and nitrogen accumulation in the grain of four cultivars of wheat. *Australian Journal of Plant Physiology*, 12: 363–375.

Bhullar, S.S. and Jenner, C.F., 1986. Effects of temperature on conversion of sucrose to starch in the developing wheat endosperm. *Australian Journal of Plant Physiology*, 12: 606–615.

Carter, T.R., Porter, J.H. and Parry, M.L., 1992. Some implications of climatic change for agriculture in Europe. *Journal of Experimental Botany*, 43(253): 1159–1167.

Chaves, M.M. and Pereira, J.S., 1992. Water stress, CO_2 and climatic change. *Journal of Experimental Botany*, 43(253): 1131–1139.

Chen, II, Shen, Z.Y. and Li, P.H., 1982. Adaptability of crop plants to high temperature stress. *Crop Science*, 22: 719–725.

Clowes, F.A.L., 1954. The promeristem and the minimal constructional centre in Grass root apices. *New Phytologist*, 53: 108–116.

Clowes, F.A.L., 1956. Localisation of nucleic acid synthesis in root meristems. *Journal of Experimental Botany*, 7 21): 307–312.

Clowes, F.A.L., 1967. Synthesis of DNA during mitosis. *Journal of Experimental Botany*, 18: 740–745.

Clowes, F.A.L., 1971. The proportions of cell that divide in root meristems of *Zea mays* L. *Annals of Botany*, 35: 349–261.

Clowes, F.A.L., 1981. The difference between open and closed meristems. *Annals of Botany*, 48: 761–767.

Clowes, F.A.L. and Hall, E.J., 1966. Meristems under continuous irradiation. *Annals of Botany*, 30: 243–251.

Cooper, J.P., 1959. Selection and population structure in *Lolium*. *Heredity*, 13: 445–459.

Cooper, J.P., 1960. Short-day and low-temperature induction in *Lolium*. *Annals of Botany*, 24 (232): 232–426.

Cooper, J.P., 1963. Species and Population Differences in Climatic Response. In: *Environmental Control of Plant Growth*, (Ed.) L.T. Evans. Academic Press, 21: 381–403.

Cooper, P., Ho, T-II.D. and Hauptmann, R.M., 1984. Tissue specificity of the heat-shock response in maize. *Plant Physiology*, 75: 431–441.

Cooper, P. and Ho, T-II.D., 1983. Heat shock proteins in maize. *Plant Physiology*, 71: 215– 222.

Cooper, P. and Ho, T–II.D., 1987. Intracellular localization of heat shock proteins in maize. *Plant Physiology*, 84: 1197–1203.

Creber, H.M.C., M.S. and Francis, D., 1993. Effects of temperature on cell division in root meristems of natural populations of *Dactylis glomerata* of contrasting latitudinal origins. *Environmental and Experimental Botany*, 33(3): 433–442.

Dn, Ye. and Tachibana, S., 1994. Respiration and sugar content of cucumber plants. *Scientia Horticulturae*, 58(4): 289–301.

Donovan, G.R., Lee, J.W., Longhurst, T.J and Martin, P., 2005. Effect of temperature on grain growth and protein accumulation in cultured wheat ears. *Australian Journal of Plant Physiology*, 10: 445–450.

Eagles, C.F., 1967. Variation in the soluble carbohydrate content of climatic races of *Dactylis glomerata* (Cocksfoot) at different temperatures. *Annals of Botany*, 31: 645–651.

Engels, II and Marschner, II, 1990. Effect of sub-optimal root zone temperatures at varied nutrient supply and shoot meristem temperatures on growth and nutrient concentrations in maize seedlings. *Plant and Soil*, 126: 215–225.

Entz, M.H. and Fowler, D.B., 1990. Differential agronomic response of winter wheat cultivars to preanthesis environmental stress. *Crop Science*, 30: 1119–1123.

Francis, D. and Barlow, P.W., 1988. Temperature and the cell cycle. In: *Plants and Temperature*, (Eds.) S.P. Long and F.I. Woodward. Symposia of the Society for Experimental Biology, 42: 181–201, Company of Biologists.

Gardner, F.P. and Loomis, W.E., 1953. Floral induction and development in orchard grass. *Plant Physiology*, 28: 201–217.

Gonzalez-Fernandez, A., Gimenez-Martin, G. and De La Torre, C., 1971. The duration of the interphase periods at different temperatures in root tip cells. *Cytobiologie*, p. 367–371.

Gonzalez-Fernadez, A., Lopez-Saez, J.F., Moreno, and Gimenez-Martin, G., 1968. Duration of the division cycle in binucleate and mononucleate cells. *Experimental Cell Research*, 43: 255–267.

Goudriaan, J., 1992. Biosphere structure, carbon sequestering potential and the atmospheric ^{14}C carbon record. *Journal of Experimental Botany*, 43(253): 1111–1119.

Grace, J., 1988. Temperature as a determinant of plant productivity. In: *Plants and Temperature*, (Eds.) S.P. Long and F.I. Woodward. Symposia of the Society for Experimental Botany, 42: 91–107.

Grif, V.G. and Valovich, E.M., 1973. The mitotic cycle of plant cells at the minimal temperature of mitosis. *Tsitologiya*, 15: 1510–1514.

Hackett, C. and Stewart, H.E., 1969. A method for determining the position and size of lateral root primordia in the axes of roots without sectioning. *Annals of Botany*, 33: 679–682.

Hadley, P., Roberts, E.H., Summerfields, R.J. and Minchin, F.R., 1983. A quantitative model of reproductive development in cowpea (*Vigna unguiculata* L.) Walp. in relation to photoperiod and temperature and implications for screening germplasm. *Annals of Botany*, 51: 33 1–543.

Harding, S.A., Guikema, J.A. and Paulsen, G.M., 1990. Photosynthetic decline from high temperature stress during maturation of wheat. II. Interaction with source and sink processes. *Plant Physiology*, 92: 654–658.

Harlan, J.R. and Zohary, D., 1966. Distribution of wild wheats and barley. *Science*, 153: 1074–1080.

Hay, R.K.M. and Tunnicliffe Wilson, 1982. Leaf appearance and extension in field-grown winter wheat plants: the importance of soil temperature during vegetative growth. *Journal of Agricultural Science*, 99: 403–410.

Heisey, W.M., 1953. Growth and development of species and hybrids of Poa under controlled temperatures. *American Journal of Botany*, 40: 205–221.

Howard, A. and Pelc, S., 1953. Synthesis of deoxyribonucleic acid in normal and irradiated cells and its relation to chromosome breakage. *Heridity Supplement*, 6: 261–273.

Hughes, M.K. Schweingruber, F.II., Cartwright, D. and Kelly, P.M., 1984. July–August temperature at Edinburh between 1721 and 1975 from tree-ring density and width data. *Nature*, 308: 341–344.

Kalra, N., Chakraborty, D., Sharma, A., Rai, H.K., Jolly, M., Chander, S., Kumar, P.R., Bhadraray, S., Barman, D., Mittal, R.B., Lal, M. and Sehgal, M., 2008. Effect of increasing temperature on yield on some winter crops in North-east India. *Current Science*, 94: 82–88.

Kent, N.L., 1975. Wheat of the world. In: *Technology of Cereals*, Second Edition, Pergamon Press, 4: 74–95.

Knight, R., 1960. The growth of Cocksfoot (*Dactylis glomerata* L.) under spaced plant and sward conditions. *Australian Journal of Agricultural Research*, 11: 457–472.

Korner, Ch. and Reinhardt, H., 1987. Dry matter partitioning and the root length/leaf area ratios in herbaceous plants with diverse altitudinal distribution. *Oecologia*, 74: 411–418.

Krishnan, M., Nguyen, H.T. and Burke, J.J., 1989. Heat shock protein synthesis and thermal tolerance in wheat *Plant Physiology*, 90: 140–145.

Landsberg, J.J., 1974. Apple fruit bud development and growth: analysis and an empirical model. *Annals of Botany*, 38: 1013–1023.

Lange, O.L., Nobel, P.S., Osmond, C.B. and Ziegler, II., 1981. *Encyclopaedia of Plant Physiology*. Springer-Verlag, Berlin, Heidelberg, New York, 12A: 279–329.

Lande, H.M., 1953. The nature of summer dormancy in perennial grasses. *Botanical Gazette*, 114: 284–292.

Liang, D.R. and Fischer, R.A., 1977. Adaptation of semi-dwarf cultivars to rainfed conditions. *Euphytica*, 26: 129–139.

Lin, C.Y., Roberts, J.K. and Key, J.L., 1984. Acquisition of thermotolerance in soybean seedlings. *Plant Physiology*, 74: 152–160.

Lopez-Saez, J.F. Gimenez-Martin, G. and Gonzalez-Fernandez, A., 1966. Duration of the cell division cycle and its dependence on temperature. *Zeltschrift Fur Zellfurschung*, 75: 591–600.

Lopez-Saez, J.F. Gonzalez-Bernandez, F., Gonzalez-Fernandez, A. and Garcia-Ferrero, G., 1969. Effect of temperature and oxygen tension on root growth, cell cycle and cell elongation. *Protoplasma*, 67: 213–221.

Lyndon, R.F., 1979. The cellular basis of apical differentiation. In: *Differentiation and Control of Development in Plants: Potential for Chemical Modification*, (Ed.) E.C. George. British Plant Growth Regulator Group, Monograph 3, BPGRG, Wantage.

Marilyn, G. and Intyre, H.C.H., 1990. The effect of photoperiod and temperature on growth and frost resistance of winter rye root systems. *Physiologia Plantarum*, 79: 519–525.

Martinic, Z., 1973. Vernalization and photoperiodism of common wheat as related to the general and specific adaptability of varieties. In: *Plant Responses to Climatic Factors*, (Ed.) R.O. Slatyer. UNESCO, Paris, pp. 153–164.

Miglietta, F. and Porter, J.R., 1992. The effects of climatic change on development in wheat: Analysis and modelling. *Journal of Experimental Botany*, 43(252): 1147–1158.

Natchwey, D.S. and Cameron, I.L., 1968. Cell cycle analysis. In: *Methods in Physiology,* (Ed.) D.M. Prescott. Academic Press, New York, 3: 213–259.

Nobel, P.S., 1988. Principles underlying the prediction of temperature in plants, with special reference to desert succulents. In: *Plants and Temperature,* (Eds.) S.F. Long and F.I. Woodward. Symposia of the Society for Experimental Botany, 42: 1–23.

Ong, C.K., 1983a. Response to temperature in a stand of pearl millet (*Pennisetum typhoides* S. and H.). I. Vegetative development. *Journal of Experimental Botany,* 34: 322–326.

Ong, C.K., 1983b. Response to temperature in a stand of pearl millet (*Pennisetum typhoides* S and H.). 4. Extension of individual leaves. *Journal of Experimental Botany,* 34: 1731–1739.

Ongham, H.J. and Howarth, C.J., 1988. Temperature shock proteins in plants. *Society for Experimental Biology,* 259–280.

Ortiz, R., Sayre, K.D., Govaertes, B., Gupta, R. and Subbarao, G.V., 2008. Climate change: Can wheat beat the heat. *Agriculture, Ecosystem and Environment,* 126: 46–58.

Ougham, H.J. and Stoddart, J.I., 1986. Synthesis of heat-shock protein and acquisition of thermotolerance in high-temperature tolerant and high-temperature susceptible lines of *Sorghum. Plant Science,* 44: 163–167.

Pahlavanian, A.M. and Silk, W.K., 1988. Effect of temperature on spatial and temporal aspects of growth in the primary maize root. *Plant Physiology,* 87: 529–532.

Powell, M.J., Davies, M.S. and Francis, D., 1986a. The influence of zinc on the cell cycle in the root meristem of a zinc-tolerant and a non-tolerant cultivar of *Festuca ruba* L. *New Phytologist,* 102: 419–428.

Powell, M.J., Davies, M.S. and Francis, D., 1986b. Effects of zinc on cell nucleolar size and on RNA and protein content in the root meristem of a Zn-tolerant and a non-tolerant cultivar of *Festuca rubra* L. *New Phyologist,* 104: 671–679.

Powell, M.J., Davies, M.S. and Francis, D., 1988. Effects of zinc on meristem size and proximity of root hairs and xylem elements to the root tip in a Zn-tolerant and non-tolerant cultivar of *Festuca rubra* L. *Annals of Botany,* 61: 723–726.

Prince, H.J., 1976. Evolution of DNA content in higher plants. *The Botanical Review,* 42(1): 27–52.

Rijven, A.H.G.C., 1986. Beat inactivation of starch synthase in wheat endosperm tissue. *Plant Physiology,* 81: 448–453.

Rufty, T.W., Raper, C.D. and Jackson, W.A., 1981. Nitrogen assimilating root growth and whole plant responses of soybean to root temperature and carbon dioxide and light in the aerial environment. *New Phytologist,* 88: 607–619.

Roberts, E.H., Hadley, P. and Summerfield, R.J., 1985. Effects of temperature and photoperiod on flowering in chickpeas (*Cicer arietinum* L.). *Annals of Botany,* 55: 191–223.

Roeckner, E., 1992. Past, present and future levels of greenhouse gases in the atmosphere and model projections of related climatic changes. *Journal of Experimental Botany*, 43(253): 1097–1109.

Rosell, C.M. and Collar, C., 2009. Effect of temperature and consistency on wheat dough performance. *International Journal of Food Science and Technology*, 44: 493–502.

Russell, G., Elks, R.P., Brown, J., Milburn, G.M. and Hayter, A.M., 1982. The development and yield of autumn and spring-sown barley in south east Scotland. *Annals of Applied Biology*, 100: 167–178.

Russell, R.S., 1977. The growth and form of root systems. In: *Plant Root Systems: Their Function and Interaction with the Soil*. McGraw-Hill Book Company Limited, UK, pp. 30–61.

Saadalla, M.M., Shanahan, J.F. and Quick, J.S., 1990. Heat tolerance in winter wheat. I. Hardening and genetic effects on membrane thermostability. *Crop Science*, 30: 1243–1247.

Shanahan, J.F., Edwards, I.B., Quick, J.S. and Fenwick, J.R., 1990. Membrane thermostability and heat tolerance of spring wheat. *Crop Science*, 30(2): 247–251.

Silsbury, J.H., 1961. A study of dormancy, survival, and other characterstics in *Lolium perenne* L. I at Adelaide, S.A. *Australian Journal of Agricultural Research*, 12: 1–9.

Sinclair, T.R., 1992. Mineral nutrition and plant growth response to climate change. *Journal of Experimental Botany*, 43(253): 1141–1146.

Smith, W.K., 1978. Temperatures of desert plants: another perspective on the adaptability of leaf size. *Science*, 201: 614–616.

Sutcliffe, J., 1977. Plants and temperature, In: *Studies in Biology*. Published by Edward Arnold Limited, 41 Bedford Square, London WCIB 3 DQ, 86: 1–57.

Taylor, A.T. and Clowes, F.A.L., 1978. Temperature and the coordination of cell cycles within the root meristem of *Allium sativum* L. *New Phytologist*, 81: 671–680.

Thorn, G.N., Ford, M.A. and Watson, D.J., 1967. Effects of temperature variation at different times on growth and yield of sugar beet and barley. *Annals of Botany*, 31: 71–101.

Torrey, J.G., 1959. A chemical inhibitor of auxin-induced lateral root initiation in roots of *Pisum*. *Physiologia Plantarum*, 12: 873–887.

Van't'Hof. J., Bjerknes, C.A. and Clinton, J.H., 1978. Replicon properties of chromosomal DNA fibres and the duration of DNA synthesis of sunflower root-tip meristem cells at different temperatures. *Chromosoma*, 66: 161–171.

Warren-Wilson, J., 1966. Effect of temperature on net assimilation rate. *Annals of Botany*, 30: 753–761.

Watson, D.J., 1958. The dependence of net assimilation rate on leaf-area index. *Annals of Botany*, 22: 37–54.

Watts, W.R., 1971. Role of temperature in the regulation of leaf extension in *zea mays*. *Nature*, 229: 46–47.

Webster, P.L. and MacLeod, R.D., 1980. Characteristics of root apical meristem cell population kinetics: A review of analyses and concepts. *Environmental and Experimental Botany*, 20: 335–358.

Wiegand, C.L. and Cuellar, J.A., 1981. Duration of grain filling and kernel weight of wheat as affected by temperature. *Crop Science*, 21: 95–101.

Wimber, D.E., 1966. Duration of nuclear cycle in *Tradescantia* root tips at three temperatures as measured with tritiated thymidine. *American Journal of Botany*, 53: 21–24.

Woodward, F.I., 1987. Temperature and the distribution of plant species. In: *Plants and Temperature*, (Eds.) S.P. Long and F.I. Woodward. Symposia of the Society for Experimental Biology, pp. 59–73.

Woodward, F.I. and Friends, A.D., 1988. Controlled environment studies on the temperature responses of leaf extension in species of *Poa* with diverse altitudinal ranges. *Journal of Experimental Botany*, 39(201): 411–420.

You, L., Rosegrant, M.W., Wood, S. and Sun, D., 2009. Impact of growing season temperature on wheat productivity in China. *Agricultural and Forest Meteorology*, 149: 1009–1014.

Zeven, A.C., 1979. Polyploidy and domestication: The origin and survival of polyploids in cytotype mixtures. In: *Polyploidy: Biological Relevance*, (Ed.) Walter H. Lewis. Plenum Press, New York and London, pp. 358–407.

Chapter 11

Structure and Behaviour of Chromosomes in Radish (*Raphanus sativus* L.)

Narsinha Dayal

Cytogenetics Laboratory, Department of Botany,
Ranchi University, Ranchi

The cultivated radish, *Raphanus sativus* L. (Family: Brassicaceae) is an annual herbaceous vegetable crop grown widely in India throughout the year and the world over for a variety of economic reasons: vegetables, condiments, tongue tingling curries, 'chutneys', mouth watering pickles, salads, preserves and oil. It is one of the richest sources of free amino acids, fat, protein, carbohydrate, sugar, vitamin C, thiamine, riboflavin, nicotinic and ascorbic acids and considerable amount of minerals. It has several medicinal properties. It is a good diagestor for it cures constipation, facilitates complete evacuation and helps in chronic gastric ailments. In homeopathy it is used in neuralgic headaches, sleeplessness and chronic diarrhoea. The juice of fresh leaves in used as diuretic and laxative. The seeds are known to be peptic, expectorant, diuretic and carminative. When consumed by women, it helps in their trouble free and painless menstrual flow. Its oil constitutes an important gradient in folk medicine. It has some anticancerous properties too as it has a good source of anticarcinogens (Anonymous 1969). Some indoles found in radish activate the microsomal mixed-function oxidase activity to the point of detoxification of carcinogenic compounds. Some other constituents of this vegetable decrease carcinogenicity by altering cytochrome p-450.

Despite being one of the most important crops, radish has not been paid due attention by cytologists, geneticists and plant breeders. Yarnell (1956) first reviewed the cytogenetical studies on cruciferous vegetable crops in which he included radish also. Dayal (1986) and Kirilova and Dayal (1990) made an humble attempt to assemble all such informations till date and to present them in collective form where almost every aspect of cytogenetics of this valuable root crop is concerned. The present article deals with the informations available thus far on the structure and behaviour of chromosomes in this valuable root crop:

Number and Morphology of Chromosomes

Due to small size of chromosomes cytological studies in radish have been restricted, primarily to the chromosome counts and morphology. The somatic chromosome number in various species of Raphanus studied so far is 2n=18 (Federov, 1969). Richharia (1937) while studying the *somatic* hromosome morphology could recognize six types of chromosomes in *R. sativus* on the basis of size and primary constriction and assigned the karyotypic formula $A_4 + B_4 + C_4 + D_2 + E_2 + F_2$. According to him A, D and F are submedian, B median and C and E subterminal. But Mukherjee (1975) recognizes only three types of chromosomes in this species with the Karyotypic formula $A_2 + B_2 + C_{14}$. A is a pair of long and nearly medium chromosomes with a primary and secondary constriction and B–a pair of medium sized chromosomes with median to submedian primary constriction while C includes seven pairs of small chromosomes with median to submedian primary constriction. Intervarietal variation in karyotype has also been shown by some workers in several varieties of radish (Sharma and Dutta, 1961; Mukherjee, 1979). It has been pointed out that in *Brassiceae*, including *Raphanus*, the basic chromosome number is x = 6 from which n = 9 has originated due to secondary balance or by some other mechanism (Kamala, 1974; Kamala *et al.*, 1980; Sikka and Sharma, 1979).

Chromosome Behaviour during Meiosis

Only a few studies have been made on meiotic chromosomes in radish. There is only one report on the pachytene chromosomes in *R. raphanistrum* where all the nine chromosomes have been identified on the basis of total length, relative lengths of long and short arms and the number, size and distribution pattern of the heterochromatic segments of varying sizes flanking the either side of the centromere (Kamala, 1974). The 9th chromosome, in addition to the heterochromatic segment, has a distal dark staining spherical satellite and intercalary heterochromatic segment which is identified as the nucleolar organizer. Sutaria (1930) and latter on Maeda and Saski (1934) studied the chromosome behaviour in the pollen mother cells (pmcs) of the Indian and the Japanese radishes respectively. In the former, although meiosis is normal, occasionally a tetravalent or even a hexavalent is observed. Two or three bivalents sometimes group together in a secondary association at diakinesis. Similar observations have also been made in the Japanese radish. It is, however, the intergeneric and interspecific hybrids where chromosome behaviour at meiosis has been sufficiently studied in order to understand the chromosome homology of various genomes that characterize different species of *Brassica* and *Raphanus*. In these hybrids formation of varying number of bivalents and univalents are chracteristic (Dayal 1986).

Meiosis has been studied more systematically and extensively in autopolyploids by Japanese cytologists and by us and will be described in some detail latter on. In cochiploids of radish and their hypo- and hyperpolyploid derivatives is in similar to that known for autopolyploids of any other species.

Occurrence of autotetraploid (Dayal 1975) and interchange heterozygotes (Kirilova and Narbut, 1979) have been reported in highly homozygous inbred lines of the cultivated radish. These are in fact cases of chromosome polymorphism which probably arises in nature due to '*Karvotvoiic orthoselection*' of white (1977) and/or to the consequence of increased homozygosit? caused by prolonged inbreeding as shown in rye (Giraldez *et al.*, 1979). Stebbins (1963) opines that interchanges are found 'floating' in high frequency in allogamous populations. Besides, in homozygotes errors do take place quite often at the premeiotic mitosis (Rees and Jones, 1977). In the wild radish, *R. raphanistrurn*, Kamala *et al.* (1980) have produced artificially three trisomics for each of the 5th, 6th and 9th chromosome and a double trisomic for the 5th and 6th chromosomes and a tetrasomic for the 9th chromosome. They have also studied the chromosome behaviour at pachytene in them.

Genetic Control of Meiosis

Although any investigation of meiosis, in effect, is a study of the genetic control of the chromosome behaviour but the most important tool for understanding the nature of the genetic system regulating the chromosome pairing is perhaps provided by a comparative study of the highly homozygous inbred lines, their various hybrids and the original populations. With this view in mind meiosis in radish has been studied by us (Fadeyeva and Dayal, 1974, Dayal, 1977 a, b, 1978, 1979). Inbred lines have an increased frequency of meiotic chromosome abnormalities, particularly the presence of univalents at metaphase I and bridges and fragments at anaphase I, and high pollen sterility in comparison to their F, hybrids and the original copulation. Besides, inbred lines also show a reduced stainability with Feulgen. We have made an extensive cytological study of three highly homozygous inbred lines, LS-337/24, LS-337/25, LS-43/51, their F_1 F_2 and backcross hybrids and the original population in order to understand the genetics of chromosome behaviour. Inbred lines demonstrate undoubted inferiority to their original population in several chromosome characters at meiosis: reduced chiasma frequency at diakinesis and metaphase I, higher coefficient of chiasma terminalization, stickiness, increased frequency of univalents, occasional occurrence of polyploid and giant cells, presence of laggards, fragments and bridges at anaphase-I etc. They also vary significantly in mean chiasma frequency, its distribution and several other chromosome characters. In sharp contrast to the inbred lines, their F_1, hybrids are remarkably uniform and regular in chromosome behaviour and resemble the original population. In them there is a significant increase in mean chiasma frequency that almost touches the population level. Other chromosomal abnormalities also disappear. This may be regarded as inbreeding depression and heterosis in chromosome behaviour.

In a heterozygous and allogamous population like radish chromosome pairing, more particularly the chiasma frequency, displays a definite limit of variation and may be considered as an adaptive character. This property of the population is probably maintained by heterozygosity and outbreeding. Forced inbreeding breaks up the heterozygosity and the buffering property of the population, giving rise to several genotypes with unbalanced genetic system. The inefficient chromosome behaviour in the inbred lines may be attributed in the genetic unbalance and to the segregation of a particular homozygous combination of genes. Variation between the inbred lines in chromosome behaviour definitely shows its genotypic control. Normalization of meiosis, particularly the restoration of chiasma frequency in the F's, almost amounts to the establishment of the genotypic control of chromosome behaviour. A return to the natural mode of reproduction, characteristic for the species, restores the genetic balance as well as the normal chromosome behaviour. Thus, the chromosome behaviour is concomitant with the breeding behaviour of the species.

Meiosis is a very complicated process which requires a more intricate system of genic control. In radish dominance and complementary action of genes are operative in regulation of an optimum level of chiasma frequency. We have demonstrated that the chiasma frequency segregates like any other aspect of the phenotype. For this a cross was made between two highly homozygous inbred lines differing significantly in mean chiasma frequency, F_1's were quite uniform and had higher chiasma frequency than their parents, probably due to overdominance of genes controlling this nuclear phenotype. Interestingly, all the 12 plants which were derived from a single plant of F_1's showed a wider distribution in chiasma frequency, some of which almost touching the level of their parents, it showed a continuos nature of variation, indicating a polygenic control for this character. Back cross hybrids showed more variation than F_1's, pointing out the genotypic influence on chiasma frequency (Figure 11.1).

Homozygous genotypes produced by inbreeding of an allogamous population differ from heterozygous ones in response to the environmental fluctuations during development (Lerner, 1954). In general heterozygotes are developmentally more stable than homozygotes. This has been shown for chiasma frequency and pollen sterility in radish. From Figure 11.1 it is evident that variation in chiasma frequency induced by the same environmental fluctuation is much less between the individuals of heterozygotes than between those of homozygotes. This fact has been proved more convincingly for pollen sterility (Dayal, 1976).

Chromosome Behaviour in Autopolyploids

Although in the past there has been some isolated studies on the chromosome behaviour in the polyploid forms of radish, it has been studied in great details by Tokumasu (1961) and by us (Singh 1991, Singh 1992).

Tokumasu (1961) made an extensive cytogenetical study on colchicine induced triploids and tetraploids in order to understand the merchanism of collapse and maintenance of polyploidy in radish and to exploit it in the improvement programme of this valuable root crop. Auto triply-ploids have characteristic trivalent configurations along with bivalents and univalents in varying number, but the frequencies of bi- and trivalents are quite high. Besides, they show considerable

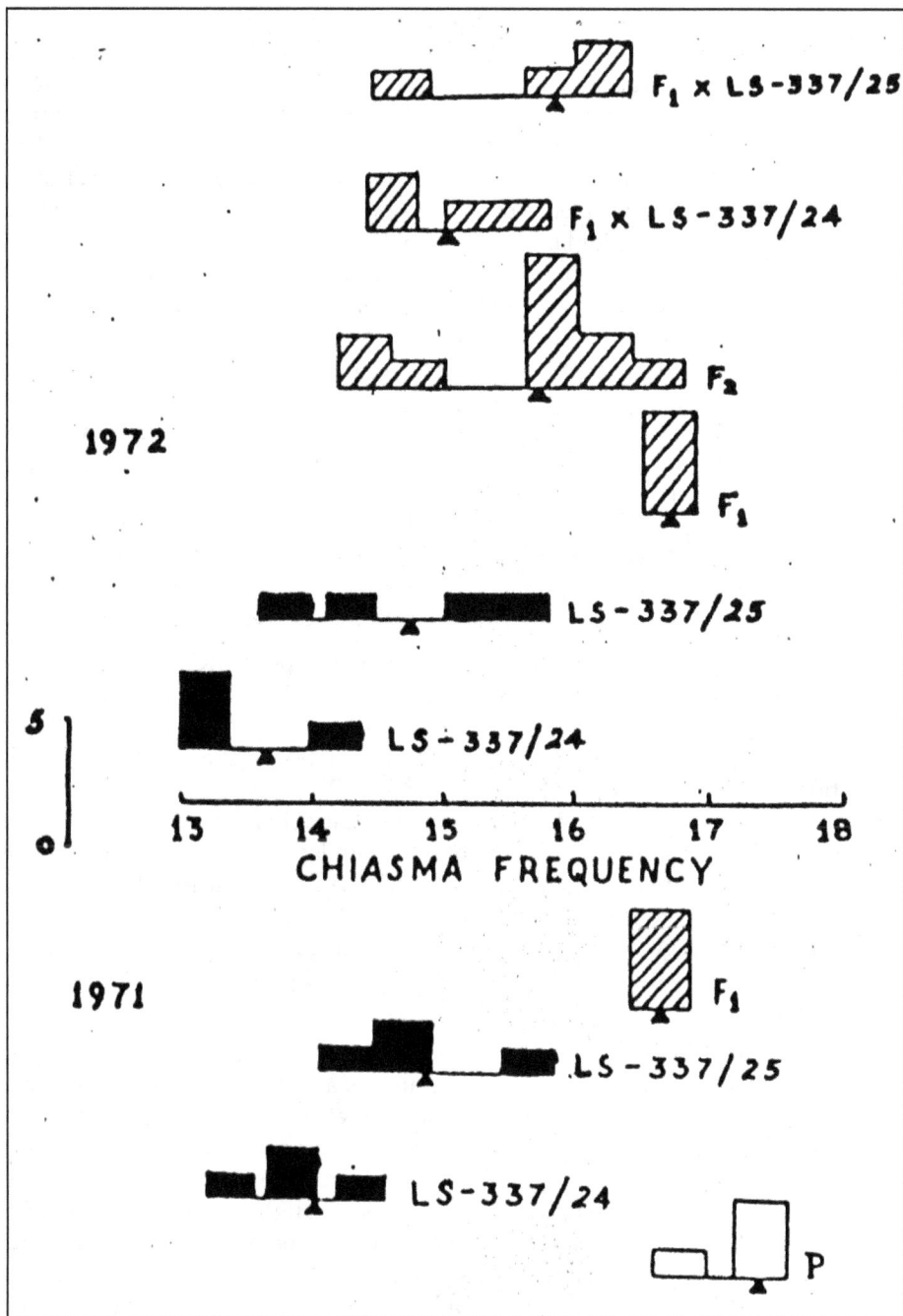

Figure 11.1: Distribution of Chiasma Frequency in Two Inbred Lines,
Namely LS-337/24 and LS-337/25, and their F_1, F_2 and Backcross Hybrids.
The horizontal scale shows chiasma frequency per cell while the vertical scale
denotes the number of plants. Arrows indicate mean chiasma frequencies.

frequency of laggards and anaphase bridges, Anaphase segregation is quite uneven, ranging from 9 to 18 chromosomes at either pole. Anaphas II is also irregular leading to the formation of monads, diads, triads, pentads, and hexads. The average number of chromosomal association per PMC is 7.1 III, 2.5 II and 0.66 I. As regards percentage chromosomal association trivalent formation occur more frequently (69.OX) than bivalents (35 per cent) and univalents (9.2 per cent). Most of the trivalents are of chain type. A considerably high frequency of chromosome abnormalities at anaphase I is a characteristic feature of triploid plants. Most of PMCs have unequal segregation of chromosomes at each pole. About 70 per cent of PMCs shows 14: 13 distribution of chromosomes at each pole while remaining ones show highly disturbed distribution. Presence of laggards and anaphase bridges are also found but in low frequency. Abnormal sporoids and pollen sterility also occur in low frequency (6.7 per cent). In autotetraploids, on the other hand, with four sets of homologues chromosomes produce five types of associations at metaphase I *viz.*, all tetravalents, one trivalent + one trivalent + one univalent, all bivalents, one bivalent + two univalents and four univalents. The different autotetraploid lines exhibit different types of meiotic chromosome configurations in varying proporations. In C_1-generation meiosis was more disturbed some of which became somewhat stabilized in C_2-generation. However, the successive generations (C_3–C_6) showed an increase in the frequency of II and IV with considerable decrease, almost to the negligible level, in that of I and III.

The average number of chromosomal association per PMC in 10 promising tetraploid line and with it ranges in C generation was 0.8I (0-3) + 9.6II (0-8) + 0.7III (0-2) + 3.4IV (0-7) and in C-generation was 0-4I (0-2) + 10.4II (3-18) + 0.2III (0-2) + 3.5IV (0-7). It clearly indicates the increase in II and IV association with a decline in I and III association in successive generations. Most of the IV were of chain or ring type. Abnormal sporoids and pollen sterility also came down sharply from 24.63 per cent in C_1 generation to 9.5 per cent in C_2 generation. A similar study of meiosis was made in two of our most promising tetraploid lines, T-87 and T-155, upto fourth generation. They showed a very high frequency (over 98 per cent) of bivalents and tetravalents in C_4 generation, although of univalents and trivalents also occasionally occurred. This indicates not only the stabilisation but also of the 'diploidization' of the meiotic chromosome behaviour in them.

It may be mentioned here that in some of our autotetraploid lines, T-87 and T-155, with high agronomical values, appearance of aneuploids is in extremely low frequency and does not hinder the maintenance of tetraploid strain. These lines are in no way inferior to their diploid counterpart in seed germination and seedling survival and have fairly good seed set. Over the years we have succeeded in improving the seed yield in our promising tetraploid lines through rigorous selection for increased frequency of bivalent and tetravalent formation as a result of which the frequency of uni and trivalent formation has been reduced or has become almost negligible. They are now meiotically well balanced having practically no meiotic chromosome abnormality and over 95 per cent pollen fertility. They have been christened as 'ND-Bijay' and 'ND-Tetra' respectively and can be commercially exploited.

References

Anonymus, 1969. *The Wealth of India.* Publication and Information Directors, CSIR, New Delhi, 8: 366–373.

Dayal, N., 1975. Cytological study of a polyploid of radish obtained as a result of inbreeding. *Curr. Sci.*, 44: 202–203.

Dayal, N., 1976. Effect of environmental fluctuations an pollen steritely in homo- and heterozygotes of radish. *Curr. Sci.*, 45: 304.

Dayal, N., 1977a. Cytogenetical studies in the inbred lines of radish (*Raphanus sativus* Var. *radicola* Pers.) and their hybrids. I. Chiasma frequency. *Cytologiax* (Tokyo), 42: 29–35.

Dayal, N., 1977b. Cytogenetical studies in the inbred lines of radish (*Raphanus sativus* Var. *radicola* Pers.) and their hybrids. II. Genetic regulation of chiasma frequency. *Cytologia* (Tokyo), 42: 273–278.

Dayal, N., 1978. Inbreeding depression and heterosis in chromosome behaviour of radish (*Raphanus sativus* L. Var *radicoia* Pears). *Curr. Sci.*, 47: 125–127.

Dayal, N., 1986. Cytogenetics of the cultivated radish. *Indian Rev. Life Sci.*, 6: 41–66,

Fadeyeva, T.S. and Dayal, N., 1974. Study of meiosis in the inbred lines of radish (*Raphanus sativus* L. Var. *radicola* (Pers.). Issleyd. *Genet.*, 5: 103–112 (In Russian).

Federov, A.A., 1969. *Chromosome Number of Flowering Plants.* Nauka, Leningrad–Moscow.

Giraldez, R., Germeno, M.C. and Orellana, J., 1979. Comparison of C-banding pattern in the chromosomes of inbred lines and open pollinated varieties of rye. *Pflanzanzuchtung,* 83: 40–48.

Kamla, T., 1974. Pachytene chromosomes in *Raphanus raphanistrum* L. *Indian J. Genet. and Plant Breed.*, 34: 44–53.

Kamla, T., Naggmani, B., Sasikala, S. and Rao, R.N.B., 1980. Cytology of aneuploids in *Raphanus. J. Cytol. and Genet.*, 15: 144–148.

Kirilova, G.A. and Dayal, N., 1990. Genetics of radish. In: *Genetics of Crop Plants,* (Eds.) T.S. Fadeyeva and L.A. Vibrureninal. Agropromizdat, Leningrad, pp. 215–239 (In Russian).

Maeda, T. and Saski, T., 1934. Chromosome behaviour in the pollen mother cells of "Chogoinadaikon' and 'Kerimadaikan', the horticulatural varieties of *Raphanus sativus* L. *Japan. J. Genet.*, 10: 76–83.

Mukherjee, P., 1975. Chromosome study as an aid in tracing the evolution of cruciferai. *Cytologia* (Tokyo), 40: 427–434.

Mukherjee, P., 1979. Karyotypic variation in nine strains of Indian radish (*Raphanus sativus* L.). *Cytologia* (Tokyo), 44: 347–352.

Rees, H. and Jones, R.N., 1977. *Chromosome Genetics.* Edward Arnold, London.

Richharia, R.H., 1937. Cytological investigations of *Raphanus, sativus. Brassica cleracea* and their F$_1$ and F$_2$ hybrids. *J. Genet.*, 34: 19–44.

Sharma, A.K. and Dutta, K.B., 1961. Interstrain differences in Karyotypes of *Raphanus sativus* L. *Indian Agr.*, 5: 151–159.

Sikka, K. and Sharma, A.K., 1979. Chromosome evolution in certain genera of Brassicaceae. *Cytologia* (Tokyo), 44: 467–473.

Singh, B.S., 1991. Studies on the cytogenetical effects of colchicine in radish (*Raphanus sativus* L.). *Ph.D. Thesis*, Ranchi University, Ranchi.

Singh, P., 1992. Cytogenetical study of the diploid, triploid and tetraploid forms of radish (*Raphanus sativus* L.). *Ph.D. Thesis*, Ranchi University, Ranchi.

Stebbins, G.L., 1963. *Variation and Evolution in Plants.* Columbia Univ. Press, New York.

Sutaria, R.N., 1930. Microsporogenesis in *Rophanus sativus* L. *J. Indian Bot. Soc.*, I: 235–256.

Tokumasu, S., 1961. The maintenance and collapse of polyploidy in the progenies of autotetraploid Japanese radishes with reference to the occurrance of aneuploid plants. *Mem. Ehime Univ., Matsuyama*, pp. 177–379.

White, M.J.D., 1977. *Modes of Speciation.* Freman and Co., San Franscisco.

Yarnell, S.H., 1956. Cytogenetics of vegetable crops. II. Cruciferae. *Bot. Rev.*, 22: 81–166.

Chapter 12

Seed Mycoflora of Umbelliferous Spices

S. Jariwala, Kanak Manjari and Bharat Rai

Centre of Advanced Study in Botany,
Banaras Hindu University, Varanasi – 221 005

ABSTRACT

The most dominant seed borne fungi reported from umbelliferous spices like, dill, ajawain, cumin coriander and fennel are *Aspergillus flavus, A.niger, A.terreus, A.sydowi, A.fumigatus, Alternaria* and *Fusatium* followed by the species of *Cladosporium, Curvularia, Penicillium, Rhizopus, Chaetomium* and *Drechsiera*. Some other fungi recorded frequently are *Epicoccum pitrpurascens, Memnonielln echinata, Stachybotrys aria* species *of Phoma, Mucor, Colletotrichum, Stemphylium* and sterile mycelia. Studies on seed treatment with few fungicides, antibiotics, culture filtrates and extracts of some plant parts have been found to reduce the seed mycoflora but increased seed germination.

Keywords: Seed mycoflora, Umbelliferous spices, Seed germination.

It is well known that the fungi associated with seeds cause their deterioration and spoilage, particularly in storage, and adversely affect the quality. Various aspects of seed mycoflora have been reviewed and discussed already by several workers (Christensen, 1956, 1973, 1980, 1991; Christensen and Lopez, 1968; Christensen and Kaufmann, 1965, 1969; Dharam Vir, 1974; Chenulu and Dharam Vir, 1979; Rai and Kanak Manjari, 1985a, 1986; Lacey *et al.*, 1991; Singh *et al.*, 1994). A number of plant diseases are also caused by seed-borne fungi which have been listed earlier by Noble and Richardson (1968) and Richardson (1979).

It has been found that losses caused by seed mycoflora in the field as well as in the storage can be reduced by chemical treatments. Fungicidal seed treatments are known to reduce the seed mycoflora and there by to improve the seed germination (Grewal and Kapoor, 1966; Mills and Wallage, 1968; Dharam Vir *et al.*, 1970; Swarup, 1970, Ellis *et al.*, 1975; Rai and Singh, 1979; Srivastava, 1984; Rai and Kanak Manjari, 1984; Shroti *et al.*, 1986; Mukhtar *et al.*, 1991, Pande and Varma, 1992; Paul and Mishra, 1992; Narain and Biswal, 1993; Shah and Jain, 1993; Klich *et al.*, 1994). The use of antibiotics to control scud mycoflora has been reported by Rai and Kanak Manjari (1986) and by Kelly and Briggs (1992).

A number of plants contain antifungal substances in their different pails (Sheik and Agnihotri, 1977; Upadhyay, 1978; Bilgrami *et al.*, 1980; Tewari and Datt, 1984; Weidenborner *et al.*, 1990). However, a little work has been done on the effects of extracts of plant parts on seed mycoflora (Pandey, 1982; Rai and Kanak Manjari, 1985b; Krishna Rao and Ratanasudhakar, 1992). A number of seed-borne fungi are capable of producing mycotoxin to cause serious diseases or physiological disorders in human and animals. A nice account on natural occurrence of mycotoxms in food has been presented by Bilgrami (1984). The culture filtrates of some dominant seed-borne fungi are reported to reduce the percentage incidence as well as the number of fungal species (Randhawa and Aulakh, 1984; Rai and Kanak Manjari, 1985c).

Although a good deal of work has been done on various aspects of seed microflora of cereals, pulses, oil seeds and vegetables a delaited investigation on several aspects of seed pathology of spices still warrants altention of the workers. Spices are substances which are used to season and flavour various food preparations. India is considered to be the traditional home of spices from ancient time. The aroma of spices is due to volatile essential oils, pungency due to alkaloid-like substances and colour due to fat or water soluble pigments (Singh *et al.*, 1994). In the present communication the authors have attempted to review the work on various aspects of seed mycoflora of some umbelliferous spices *viz.*, **dill** (*Anethum sowa* Kurs.), Ajawain (Tracyspermum *amuni* (L.) sprague ex Turrill), coriander (*Coriandrum sativum* L.), Cumin (*Cuminum cyminum* L.) and fennel (*Foeniculum vulgare* Mill.).

The seed mycoflora has been studied by the workers either from freshly collected samples or from storage, at various time intervals, or from the both. The mycoflora has been isolated generally from unsterilized and surface sterilized seeds by a combination of normally two techniques *i.e.* the 'agar plate method' and the 'blotter method' as recommended by the ISTA (1986). The latter method has been found better than the former by most of the workers.

Seed Mycoflora of the Spices

The first preliminary report on seed mycoflora of any spice *i.e.* **dill** was published by Janardhan and Ganguly (1963) while studying fungal flora of seeds of some important medicinal plants. They observed that the fungal flora of dill seeds predominantly consisted of species *of Alternaria, Aspergillus* and *Penicillium* (Table 12.1). *Cladosporium* and *Mucor* species were less prevalent and those of *Chaetomium* were rarely recorded. A reduction in the fungal flora and bacteria was observed by surface sterilization of seeds with mercuric cloride and also by seed treatment with Ceresan.

Swarup and Mathur (1972) studied seed mycoflora of ajawain, coriander, cumin, dill and fennel and isolated species of *Alternaria, Aspergillus Fusarium, Penicillium, Chaetomium, Curvularia lunata, Memnoniella grisea, Memnoiella echinata, Rhizopus arrhizus, Rhizoctonia bataticola* and *Siemphylium botryosum* from seed surface of freshly collected spices. The fungi isolated from the interior of seed were almost all the species of *Penicillium* and *Fusarium* (Table 12.1).

The isolates from five umbelliferous spice seeds, collected from six different localities, did not show significant difference amongst the list of fungi metioned above. However, there was some variability in the range of total number of organisms isolated from seed surface and from within the tissues of the seed (indicated in parenthesis), *viz.*, ammi 17-23 (3-9); coriander 16-21 (4-7); cumin 14-21(3-5); dill 11-21 (1-5); fennel 13-23 (2-7).

Gupta and Neergaard (1970) detected clamydospores of *Proiimtyces macrosporus* in Indian seed lots of corinader. Richardson (1979) has listed some seed-borne pathogenic fungi from Indian lots of coriander which include *Alternaria poonesis,Colletotrichum capsici, Gloeosporium achaenhcola* and *Protomyces macrosporus* (Table 12.1).

Anahosur *et al.* (1972) isolated species of *Allernaria, Aspergillus, Candida, Cercospora, Cladosporium, Curvularia, Helminthosporium, Memnoniella, Monilia, Mucor, Penicillium, Rhizopus,* and nonsporulating unidentified siptate mycelia from fennel seed (Table 12.1). They reported species of *Cercospora, Aspergillus, Mucor, Rhizopus* and *Alternaria* to be internally seed-borne fungi. The former caused necrosis of the leaves as well as petioles where as the latter four caused rotting of seed as well as of the seeding.

Prasad (1979, 1980 a, b) studied some aspects of seed mycoflora of coriander. He concluded that most of the seed-borne fungi showed high pectinolytic activity. Cellulolytic and amylolytic activity was shown by all the recorded fungi. In a separate study he found that the seed fungi were responsible for decreasing pH, dry weight, and oil and nitrogen content of the stored seeds but incresing the electrolyte leakage. Most of the seed mycoflora recorded by him were species of *Aspergillus, Chaetamium, Curvularia, Cladosporium, Fusarium,* and a few other fungi including Trichoderma (Table 12.1).

Prasad (1980c) further studied the incidence of preharvest endophytic seed mycoflora of some umbelliferous spices *viz.*, fennel, ajawain, dill and cumin and reported that the seeds were contaminated infernally with fungi mostly belonging to Dematiaceae. The fungi recorded were the species of *Alternaria, Curvularia, Drechslera, Cladosporium, Epicoccum purpurascens, Fusarium moniliforme, Memmmiella echinata* and *Stachybotrys atra* (Table 12.1).

Narayan and Prasad (1981) studied the succession of seed mycoflora on stored fennel and recorded altogether 45 species from fresh to three years of storage. Minimum number of fungi was recorded on fresh seeds increasing gradually unto the second year. *Aspergillus,* particularly *A. flavus* and *A. nididans,* invaded the seeds during the first year of storage and their frequency increased during successive years. Fungi with dark coloured mycelium were noticed on fresh seeds whereas *Chaetomium* sp.,

Table 12.1: Seed Mycoflora of Umbelliferous Spices

Name of Spices	Seed Mycoflora	References
Dill	*Alternaria, Aspergillus, Pencillium, Cladosporium,* anardhanan and Ganguly (1963) *Mucor* and *Chaetomium*	Janardhana and Ganguly (1963)
Coriander	*Protomyces macrosporus, Alternaria poonesis, Colldotrichum capsici, Glocosporium achaeniicola, Botrytis cinerea. Glomerella Cingulata, phoma glomerata, Myrothecium, state of Nectria bacteridiodes, Periconia atra, Coleophoma empetri. Epicoccum Purpurascens,Nigrospora sphaerica, Alternauia, Aspergillus,curvularia, Chaetomium, Cladosporium, Drechslera, Fusarium* and *Trichoderma.*	Gupta and Neergaard (1970), Richardson (1979), Prasad (1979,1980a,b). Srivastava 1984).
Fennel (Externally seed-borne Fungi)	*Cladosporium cladosporioides, Cephaliphora irregularis, Cunninghamella echinulata, Epicoccum purpurascens, Sadasivania gresia, Stachybotrys atra, Syncephalastrum racemosum, Thielavia terricola, Alternaria, Aspergillus, Botryotrichum, Candida, Cercospora, Chaetomium, Cladosporium, Curvularia, Fusarium, Gleocladium, Helminthosporium Macrophoma, Membobielia, Monilia, Mucor, Penicillium, Rhizopus, Sclerotium,* and *Sterile mycelia.*	Anahosur et al. (1972), Narayan and Prasad (1981).
Internally seed-borne	*Cercospora, Aspergillus, Mucor, Rhizpus* and *Alternaria*	Anahosur et al. (1972)
Coriander and carum (ajawain) from non-sterilized seeds	*Aspergillus flavus, A. niger, Chaetomium aureum, C. convolutum, Curvularia pallescens, Penicillium Stecki, Rhizopus nigricans.*	Manohachary and Khalis (1983)
From surface-sterilized seeds	*Alternaria alternata, Aspergillus flavus, A. fumigatus, A. niger, Chaetomium aureum, C.convolutum, Chaetomella raphigera Cladosporium herbarum, Curvulauia maculens, C. pallescens, Drechslera halodes, D. specifer, Fusarium semitectum, Membobiella echinata, Penicillium Stecki, Phoma nebulosa, Stachybotrys atra* and *Trichoderma viride.*	Manohachary and khalis (1983)
Coriander and fennel	*Alternaria alternate, Asperillus flavus, A.fumigatus, A.nigeer, A.ochraceous, Chaetontium globosum, Curvularia lunata, C.pallescens, Fusarium Moniliforme, Helminthosporium spp. Memnoniella echiata, Penicillium oxalicum, Rhizopus Stolonifer, Stachybotrys atra* and *Trichurus* spp.	Prasad et al. (1984)
Coriander and fennel	*Rhizopus stolonifer, Aspergillus fumigatus* and *Fusarium oxysporum.*	Srivastava and Chandra (1985)
Ajwain, Coriander, cumin, dill and fennel From seed surface :	*Alternaria alternate, A. tenuissimam, Aspergillus amstelodami, A. awamori, A. flavus, A.nidulans, A.niger, A.oryzae, A.ruber, A.sydowi, A.terrus, A.unguis, Chaetomium globosum, Curvularia Lunata, Fusarium bulbigenum, F.moniliforme, Masoniella grisea, Memnoniella echinata, Penicillium capsulatum, P.oxalicum, Rhizopus arrhizus, Rhizoctonia bataicola* and *Stemphylium botryosum.*	Swarup and Mathur (1972)
From interior of seed :	Penicillium and Fusarium	Swarup and Mathur (1972)
Fennel, ajawain, dill and cumin	*Epicoccum purpurascens, Fusarium moniliforme, Memnoniella echinata, Stachybotrys atra, Alternaria, Curvularia, Drechslera* and *Cladosporium.*	Prasad (1980c)

Mucorales, *Botryatrichum* sp., *Cephaliophora irregularis* and *Fusarium roseum* appealed later (Table 12.1).

Seed mycoflora of coriander and carum was studied by Manoharachary and Khalis (1983). They isolated 40 fungal species of 23 genera. The surface sterilized seeds yields more fungal species than non sterilized ones and the reason was ascribed to the fungi like *Rhizopus nigricans, Aspergillus flavus, A. niger* and *Penicillium stecki* (Table 12.1) which perdominated nonsterilized seeds and grew faster, suppressing growth of other fungi. A few fungi like species of *Curvularia, Cladosporium, Drechslera, Fusarium* and *Phoma* were noticed persistently in surface sterilized seeds and also at the seedling stage and were thought to be potential seed-borne pathogens.

Srivastava (1984) studied seed-borne fungi from 40 Indian seed lots of coriander of 6 different states and recorded 35 fungi *viz.*, *Botrytis cinarea, Collectotrichum capsici, Glomerella cingulata, Epicoccum purpurascens, Nigrospora sphaerica Cladosporium sphaerospermum*, species of *Aspergillus, Alternaria, Curvularia, Drechslcra* and *Fusarium* causing varying degree of seed and seedling mortalities (Table 12.1).

Prasad *et al.* (1984) recorded 15 fungal species from seeds of coriander and fennel (Table 12.1). These were *Alternaria alternata, Aspergillus flavus, A. fumigatus, A. niger, A. ochraceous, Chaetomium globosum, Curvularia lunata, C. pallescens, Fusarium moniliforme, Helminthosporium* spp., *Memnoniella echinata, Pencillium oxalicum, Rhizopus stolonifer, Stachybotrys atra* and *Trichurus* spp. *A. flavus* was found to be the most dominant one. Srivastava and Chandra (1985) found that *Rhizopus stolonifer, Aspergillus fumigatus* and *Fusarium oxysporum* were highly pathogenic to coriander and fennel seeds causing their pre and post-emergence mortality.

Rai and Kanak Manjari, (1985a), Kanak Manjari (1986) have studied several aspects of seed mycoflora of fennel and ammi (Ajawain) in relation to storage period. The number of fungal species isolated from unsterilized and surface sterilized stored seeds by two (Agar plate and Blotter) methods were 47/32 and 39/24 from fennel and ammi respectively. It was observed that majority of the mycoflora isolated from the seeds of fennel and ammi were common to each oilier (Table 12.2). The dominant fungal species isolated throughout the year from unsterilized seeds of both the spices were *Alternaria alternata, Aspergillus flavus, A. luchuensis, A. niger, A.sydowi, A. terreus, Cladosporium cladosporioides, Curvularia lunata, Drechslera* state of *Cochliobolus spicifer, Penicillium citrinum, Rhizopus nigricans* and white sterile mycelia (Table 12.2). However, the dominant fungal species recorded from surface sterilized seeds of both the spices were *Aspergillus flavus, A. niger* and *Rhizopus nigricans*. It may be concluded, therefore, that these species are important seed deteriorating fungi of the test spices.

High percentage incidence of Asperigilli, Penicillin and *Rhizopus nigricans* was recorded on agar plates whereas that of *Alternaria alternata, Chaetomium bostrychodes, Cladosporium cladosporioides, Curvularia lunata* and *Drechslera* state of *Cochliobolus spicifer* on blotter wads. It was also observed in case of all the spices that rainy and summer season were most favourable for high incidence of *Aspergillus* spp., *Alternaria alternata* and *Rhizopus nigricans* whereas *Cladosporium cladosporioides Curvalaria lunata, Drechslera* state of *Chochliobolus spicifer* and *Penicillium citrinum* showed their maximum percentage of incidence in winter.

Table 12.2: Per cent Frequency of Seed Mycoflora Associated with Unsterilized and Sterilized Seeds of Stored Spices (July to June)

Name of Species	F. vulgare		T. ammi	
	US	S	US	S
Acremonium vitis	–	–	–	4
Altemaria alternata	88	40	84	25
A. brassicicola	–	4	–	–
A. longipes	–	–	4	–
A. raphani	4	–	–	–
Asperigillus Candidus	36	8	32	12
A. flavus	100	80	100	72
A. fumigatus	8	–	12	4
A. humicola	12	4	4	–
A. luchuensis	40	4	52	8
A. nidulans	32	12	25	4
A. niger	100	92	100	84
A. sulphureus	12	–	8	–
A. sydowi	60	28	64	25
A. terreus	76	28	80	1 6
Aureobasidium pullulans	4	–	12	–
Chaetomium bostychodes	56	50	44	16
Cladosporium cladosporoides	100	60	100	32
Cunninghamella ehinulata	16	8	–	8
Curvularia clavata	32	–	16	–
C.lunata	92	52	100	52
Drechslera halodes	4	–	–	–
D. papendorfi	–	–	4	–
D.state of cochliobolus spicifer	100	60	100	32
Epicoccum purpurascens	28	8	25	–
Fusarium oxysporum	–	–	4	–
F. semitectum	8	4	12	–
Gilmamella humicola	–	–	4	–
Graphium bulbicola	–	4	–	–
Humicola grisea	4	–	4	–
Memnoniella echinata	40	12	20	4
Mortierella sp.	4	–	–	–
Mucor sp.	8	–	–	–
Myrothecium sp.	4	–	–	–

Contd...

Table 12.2–Contd...

Name of Species	F. vulgare		T. ammi	
	US	S	US	S
Nigrospora sphoeriva	8	–	8	–
Pencillium citrinum	52	40	100	44
P. rubrum	12	4	–	–
P. rugulosum	12	4	–	–
Periconia hispidula	–	4	–	4
Rhizopus nigricans	100	92	100	92
Scopulariopsis brevicaulis	–	–	8	–
Spicaria divaricata	24	8	28	4
Stachybotrys atra	20	4	20	–
Stemphylium verruculosum	–	–	4	–
Syncephalastrum racemosum	16	–	8	–
Thielavia terricola	4	8	4	–
Torula herbarum	4	–	–	–
Trichoderma sp.	8	8	8	12
Ulocladium botrytis	8	–	4	–
Dark sterile mycelia	32	16	36	8
White sterile mycelia	60	50	84	52
Yellow sterile mycelia	20	4	25	18
Total no. of the species 52	42	31	39	24

US: Unsterilized seed 1–20 per cent = Rare.

S: Surface sterilized seeds 21–40 per cent = Sub-dominant.

–: Absent 41–100 per cent–Dominant.

Effects of Antibiotics and Fungicides on Seed Mycoflora

The antibiotic are biologically very active and degraded in nature and thus they normally do not posts threat for human health hazards as compared to the fungicides. Though a considerable work has been done on the effects of antibiotics against seed-borne mycoflora of several crops, very little attention has been paid on this aspect with regard to spices. Rai and Kanak Manjari (1986) have studied the effects of griseofulvin and mycostatin on seed mycoflora of ajawain and fennel at three different concentrations *viz.*, 50, 100 and 200 ppm. Both the antibiotics were effective in combating the seed mycoflora. The effect became more pronounced with the increasing concentration of the antibiotic. Griscofulvin was more effective than mycostatin in reducing the mycoflora. A few fungi which were noted to be somewhat tolerant against griseofulvin were *Aspergillus niger, Rhizopus nigricans* and *Alternaria alternata* whereas other species of *Aspergillus* and rest of the fungi were most susceptible. In case of Mycostatin *A. alternata, A. niger, Cladosporium cladosporioides, Curvularia lunata* and *Fusarium semiteclum* were noted to be, more or less, tolerent, while other species of *Aspergillus* alongwith some other were most susceptible.

A little information is available on fungicidal seed treatment of spices to control seed mycoflora Swarup (1970) studied the effect of fungicides on the storage of seeds of ajawain, coriander, cumin and fennel and reported that all 14 fungicides used considerably reduced the percentage of moulded seeds. Anahosur *et al.* (1972) investigated the control of seed mycoflora of fennel by using ten fungicides and reported that sulphur followed by Hexasan, Agrosan G N and Ceresan effectively controlled the associated seed mycoflora.

Srivastava (1984) observed elimination of almost all the seed fungi of coriander after seed treatment with 0.3 per cent Dithane M-45 and Thiride 75-D besides improving the seed germinability. Bavistin (0.4 per cent) also controlled all fungi except *Alternaria alternata* which was present on a few seeds. Some fungi like *A. alternata, Aspergillus flavus* and Curvularia lunata were recorded in very low count in the seeds treated with 0.4 per cent Ceresan dry.

Rai and Kanak Manjari (1984), Kanak Manjari (1986) studied the effect of different concentration (0.1, 0.3 and 0.5 per cent) of four- fungicides *viz.*, Bavistin, Dithane M-45, Agrosan G N and PCNB against seed mycoflora of fennel and ajawain. All the used fungicides wen: found to be effective in redacting the seed mycoflora qualitatively as well as quantitatively at each concentration. Dithane M-45 (0.5 per cent) was proved to be the best fungicide controlling seed mycoflora and improving the seed germinability. *Alternaria alternata, A. flavus* and *A. niger* showed somewhat tolerance against all the fungicides. The only exception was noted for *A. niger* in case of Bavistin on fennel when it got completely eliminated. *Curvulria lunata, C. clavata* and *Chaetomium* sp. were observed to be most susceptible to the fungicides used.

Effect of Extracts of Plant Parts on Seed Mycoflora

There is paucity of information on the effects of extracts of plant parts on seed mycoflora of spices. Srivastava (1984) reported that the aqueous extract (100 per cent) of *Canabis sativus* was effective in controlling seed-borne fungi of coriander. However, the seed treatment adversely affected the seed germination even at low concentration.

Rai and Kanak Manjari, (1985b) have studied the effects of 10, 20 and 40 per cent aqueous bulb extract of *Allium cepa,* leaf extract of *Argemone mexicana* and *Ocimum sanctum* and oil cake extract of *Azadirachta indica* on seed mycoflora of fennel and ajawain. All the extracts showed antifungal activity against the seed mycoflora and considerably reduced the number as well as percentage incidence of the fungal species. However, the exact nature of the active principles of the extracts responsible for reduction was not identified. Amongst all the extracts the leaf extract of *A. mexicana* was found to be most effective in reducing the seed fungi of all the selected spices although the extracts of *A.sativum, O.sanctum* and *A.indica* also showed pronounced effect for controlling seed mycoflora. Out of all the seed fungi *Alternaria alternata, Aspergillus flavus, A. niger* and *Cladosporium cladosporioides* were observed to be tolerant to some extent against the all extracts.

Mycotoxins Producing Seed-Borne Fungi

There are only a few reports on mycotoxin producing fungi on spices (Seenappa, 1970; Scott and Kennedy, 1973; Fannigan and Hui, 1976; Singh, 1984 and Prasad *et al.*, 1984). According to Prasad *et al.* (1984) out of 60 isolates of *A.flavus*, 35 per cent were toxigenic which elaborated aflatoxin B_1. Three isolates obtained from fennel produced aflatoxin B_z also. Quantitative estimation oflatoxin B_1 showed that the isolates from fennel were more toxigenic. Five out of 7 samples of fennel yielded 10-43 ppb aflatoxin B_1 whereas only 3 samples of coriander were aflatoxin positive (12-20 ppb).

Effects of Fungal Metabolites on Seed Mycoflora

Rai and Kanak Manjari (1985c) have studied the effect of culture filtrates of 10 dominant seed fungi of ajawain and fennel *viz.*, *Alternaria alternata, Aspergillus flavus, A.niger, A. sydawi, A. terreus. Cladosporium cladosporioides, Curvularia lunata, Drechslera* state of *Cochliobolus spicifer; Penicillium citrinum* and *Rhizopus nigricans* on their seed mycoflora. It was found that most of the fungal culture filtrates, more or less, reduced the percentage incidence as well as the number of fungal species. A majority of the dominant seed mycoflora were found to be less affected most of the culture filtrates. *Cladoporium cladosporoides* and *C. lunata* were least affected by all the filtrates. In some cases most of the filtrates showed stimulatory effect on few fungi like *Chaetomium bostrychodes* and *A. sydowi* on fennel and *Cladosporium cladosporoidies* on ajawain. However, in case of some filtrates several numerically unimportant species got completely eliminated.

Effect of Seed Treatment and Seed Mycoflora on Germination

Several workers have studied the effects of seed treatments on seed mycoflora and on seed germination of seeds of umbelliferous spices. Janardhanan and Ganguly (1963) observed that dill seeds germinated better at temperature below 32°C. It was also seen that the surface sterilization of seeds with 0.1 per cent mercuric chloride or seed treatment with dry Ceresan did not appreciably improve the seed germination. Swarup and Tandon (1971) studied the effects of microorganisms on the germination of seeds of umbelliferous spices stored at diferent temperatures and relative humidites and found that the micro organisms reduced the seed viability of ajawain, coriander, cumin and funnel to zero level within shortest period of storage at different temperatures and relative humidities. Prasad (1980b) noticed that seed mycoflora of coriander caused maximum electrical conductivity and lowest seed germination whereas the seed free from mycoflora showed minimum conductivity and highest seed germination. A reduction in seed germination of treated spices with fungal inoculants was reported by Manoharachary and Khalis (1983) when they oberved germination of only 16 per cent seeds of coriander inoculated with a fungal mixture against 50-70 per cent germination of healthy seeds. Howerver, carum seeds inoculated with a fungal mixture showed 45 per cent germination against 57 per cent healthy seeds.

Srivastava (1984) reported that although the seed treatment of coriander with aqueous extract of *Canabis sativus* effactively controlled the seed-borne fungi, it suppressed the seed germination completely and even its low concentrations did not allow seed germination. It was observed by Srivastava and Chandra (1985) that majority of the fungal species associated with quiescent and germinating seeds of coriander and fennel adversely affected seed germination. However, a number of fungi *viz..*, *Chaetominm globosum*, *Penicillium* sp., *Geotrichum* sp., *Macrophomina phaseolina* and *Alternaria alternata* increased the germination percentage and improved the vigour of the seedlings survied. It was reported by Rai and Kanak Mangari (1986) that a gradual reduction in germination of unsterilized and sterilized seeds of ajawain occured as the period of storage was prolonged.

Thind *et al.* (1977) studied the effect of four antibiotics of the seed germination of coriander and reported that low concentrations (10-50ppm) of nystatin, demeclocycline hydrochloride and tetracycline hydrochloried promoted the seed germination. Concentration higher than 75 ppm mostly inhibited the germination in all cases. Rai and Kanak Manjari (1986) reported that seed treatment of ajawain and fennel with 50-100 ppm concentration of griseofulvin and mycostatin did not show marked effect on seed germination. However, both the antibiotics showed phytotoxic behaviour on seed germination at high concentration (200 ppm). Swarup (1970) observed that seed treatment with 14 fungicides resulted in an increase in germination and seedling emergence of ajawain, coriander and cumin. However, in case of fennel all the fungicides used except Agrosan GN showed an adverse effect on germination as well as emergence. The effect of four fungicides *viz.*, Bavistin, Dithane M-45, Agrosan GN and PCNB at 0.1. 0.3 and 0.5 per cent concentrations on seed germination of ajawain and fennel was studied by Rai and Kanak Manjari (1984). It was found that the germination of ail the treated seeds was higher than that of untreated control ones except Agrosan GN which adversely affected seed germination. Dithane M-45 was proved to be the most ettective for controlling seed mycoflora and improving the seed germination.

A reduction in seed germination of some umbelliferous spices *viz.*, fennel, coriander and cumin due to fungal toxin have been reported by Tandon, Dwivedi *et al.* (1985). It was pointed out that the reduction in percentage seed germination was more pronounced in culture filtrates of *Aspergillus flavus*. It was observed by Rai and Kanak Manjari (1985c) that the culture filtrates of ten dominant seed mycoflora (listed under the title "effects of fungal metabolites on sued mycoflora" reduced the seed germination of ajawain and fennel. The authors further reported that seed treatment of ajawain and fennel with bulb extract of *Allium sativum* and oil cake extract of *Azadirachta indica* promoted the germination at each concentration (10, 20 and 40 per cent). However, the leaf extract of *Argemone mexicana* reduced the seed germination at 40 per cent.

Conclusion

Considerable information is available now on seed mycoflora of some umbeliferous spices and it is apparent that majority of the seed-borne fungi are common to those reported on other kind of seeds. However, there is little information on internally seed-borne fungi of the spices. Efforts should be made, therefore, to isolate internally seed-borne fungi, and to study histopathology of infected seeds. The mode of survival and transmission of pathogenic forms should also be investigated.

Fungicidal seed treatment to control seed mycoflora of the spices has drawn attention of several workers but this method is helpful to maintain seed viability for sowing and not for human consumption because of its harmful effect on human health. Attempt should be made to isolate some potential antagonistic microorganisms against pathogenic dominant seed-borne fungi to use them for biological control of the pathogens and seed-borne diseases.

Use of extracts and products of some plant parts having antifungal properties to control seed mycoflora of the spices is another important aspect for future research as it may not adversely affect human health. *Aspergillus flavus* and a few other mycotoxin producing fungi are frequently associated with seeds of umbelliferous spices, particularly fennel which is commonly used alter meals in India, and thus emphasis should be given to categorize mycotoxin production and to find out suitable method for their elmination.

Role of individual pathogenic fungi on deterioration and spoilage of seeds and their behaviour in association with other seed borne fungi in various combination in different environment be studied, in addition role of individual pathogenic seed-borne fungi and its ecology with particular reference to interaction with other associated microorganisms be also studied to exploit antagonize microorganism for biological control.

In tropical countries like India the method of seed storage, including those of spices, are not satisfactory to check the growing population of seed mycoflora. Moreover, the storage method suitable for one part of the country may not be suggested for other regions because of drastic variations in environmental and climatic condition in different part of the country. It is, therefore, desirable to study the effect of different environmental factors like temperature, moisture, aeration etc. on development of seed mycoflora on the spices dining preharvesting, harvesting and storage periods so that suitable method of storage after appropriate harvesting period for various regions may be suggested.

Serological and DNA hybridization tests should be followed to detect seed-borne fungi. For economic reasons it is important that high value seed be assayed by the above methods (McGee, 1995).

For proper storage, seed certification scheme and quarantine control system should be adopted. Seed health testing laboratories, research units and training centres should be established to give guidence to formers for proper storage of seeds.

References

Anahosur, K.H., Fazalnoor, K. and Narayanaswamy, 1972. Control of seed microflora of fennel (*Foeniculum vulgare* Mill.). *India. J. Agric. Sci.,* 42: 990–992.

Bilgrami, K.S., 1984. Mycotoxins in food. *J. Indian Bot. Soc.,* 63: 109–120.

Bilgrami, K.S., Misra, R.S., Sinha, K.K. and Singh, Premlata, 1980. Effect of some wild and medicinal plant extracts on inflatoxin production and growth of *Aspergillus flavus* in liquid culture. *J. Indian Bot. Soc.*, 59: 123–126.

Chenulu, V.V. and Vir, Dharam, 1979. Seed mycoflora its role in grain spoilage and production of mycotoxins. *Bull. Grain. Technol.*, 17(2): 149–157.

Christensen, C.M., 1956. Deterioration of stored grain by molds Reprinted from Wallerstein Laboratories Communications, 19: 64.

Christensen, C.M., 1973. Loss of viability in storage microflora. *Seed Sci. and Tech.*, 1: 547–562.

Christensen, C.M., 1980. Needed research on storage molds in grains seeds and their products. *Plant Dis.*, 64: 1067–1070.

Christensen, C.M. and Kaufmann, H.H., 1965. Deterioration of storage grain by fungi. *Ann. Rev. Phytopath.*, 3: 69–84.

Christensen, C.M. and Kaufmann, H.H., 1969. *Grain Storage.* University of Minnesota Extension Press, Minneapoli, p. 153.

Christensen, C.M. and Lopez, L.C., 1963. Pathology of stored seed. *Proc. Int. Seed Test. Ass.*, 28: 701–711.

Christensen, C.M., 1991. Fungi and seed quality. In: *Handbook of Applied Mycology, Vol 3: Foods and Feeds*, (Eds.) D.K. Arora, K.G. Mukerji and E.H. Murth. Marcel Dekker Inc. New York, p. 99–120.

Dharam Vir, 1974. Study on some problems associated with post-harvest: Fungal spoilage of seeds and grains. In: *Current Trends in Plant Pathology*, Prof. S.N. Das Gupta Felicitation Volume, Botany Department, Lucknow University, pp. 221–226.

Dharam Vir, Mathur, S.B. and Neergaard, P., 1970. Control of seed borne infection of *Drechslera* spp. on barley rice and oats with Dithane M-45. *Indian Phytopath.*, 23: 570–572.

Ellis, M.A., Ilyas, M.B. and Sinclair, J. B., 1975. Effected of three fungicides on internally seed-borne fungi and germination of soybean seeds. *Phytopathology*, 65(5) 553–556.

Flannigan, B. and Hui, S.C., 1976. The occurrence of inflatoxin producing strain of *Aspergillus flavus* in the mould floras of ground spices. *J. Appl. Bact.*, 41(3): 411.

Grewal, J.S. and Kapoor, S., 1966. Viability of fungicide treated wheat and barley seed in storage. *Indian Phytopath.*, 19: 179–193.

Gupta, J.S. and Neergaard, P., 1970. Deletion of chlamydospores of *Protomyces macrosporus* Ung in Indian seed lots of coriander. *Proc. Int. Seed Test. Ass.*, 35: 151–155.

International Seed Testing Association (ISTA), 1966. International rules for seed testing. *Proc. Int. Seed Test. Ass.*, 31: 1–152.

Janardhanan, K.K. and Ganguly, K.K., 1963. Fungal flora of seed of some important medicinal plants. *Indian Phytopath.*, 16: 379–381.

Kanak, Manjari, 1986. Studies on seed mycoflora of some Indian spices. *Ph. D. Thesis*, Banaras Hindu University, Varanasi.

Krishna Rao, V. and Ratrisudhakar, T., 1992. Effect of grain microflora of paddy during storage. *Indian Phytopath.*, 45(1): 55–58.

Kelly, L. and Briggs, D.E., 1992. The influence of the grain microflora on the germinative physiology of barely. *Journal of Institute of Brewing*, 98(5): 395–100.

Klich, M.A., Arthur, K.S., Lax, A.I. and Bland, J.M., 1994. Iturin A: A potential new fungicide for stored grains. *Mycopatholgia*, 127 (2) 123–127.

Lacey, J., Ramakrishna, N. and Hamex, A., 1991. Grain fungi. In: *Handbook of Applied Mycology, Vol. 3: Food and Feeds*, (Eds.) D.K. Arora, K.G. Mukerji and E.H. Marth. Marcel Dekker Inc., New York, p. 121–127.

Manoharachary, C. and Khalis, N., 1983. Seed mycoflora of coriander and carum. *Indian Bot. Reptr.*, 2(1): 90–92.

McGee, D.C., 1995. Epidemiological approach to disease management through seed technology. *Ann. Rev. Phytopthol.*, 33: 445–466.

Mukhtar, J., Rath, Y.P.S. and Khan, A A., 1991. Seed-borne mycoflora of rice bean and the effect of seed treatment on seed germinability. *Bioved.*, 2(1): 67–70.

Mills, J.T. and Wallage, H.A.H., 1968. Deterioration of selective action of fungicides on the microflora of barley seeds. *Can. J. Pl. Sci.*, 48: 587–594.

Narain, A. and Biswas, G., 1993. Seed-borne fungal diseases of few legume crops and their management. In: *6th International Congress of Plant Pathology*, Montrial, Canada, pp. 261.

Narayan, N. and Prasad, B.K. Successonal studies of seed mycoflora of stored fennel. *Acta Botanica Indica*, 9: 57 59.

Noble, M. and Richardson, M.J., 1968. An Annotated list of seed-borne disease. *Phytopathological Papers*, Common W Mycol Inst., Kew, pp. 191.

Pande, A. and Varma, K.V.R., 1992. Seed-borne fungi of pigeon pea their pathogenicity and the fungicidal control. *Biovigyanum*, 18(1): 33–38.

Pandey, K.N., 1982. Antifungal activity of some medicinal plants on stored seeds of *Eleuisine coracana. Indian Phytopath.*, 35: 499–501.

Paul, M.C. and Mishra, R.R., 1992. Studies on seed mycoflora. *Crop Research (Hisar)*, S Supplement: 225–232.

Prasad, B.K., 1979. Enzymic studies of seed-borne fungi of coriander. *Indian Phytopath.*, 32: 92–94.

Prasad, B.K., 1980a. Influence of seed mycoflora on electrical conductivity and seed germination of coriander. *Indian Phytopath.*, 33: 138.

Prasad, B.K., 1980b. Fungus induced physiochemical change in stored coriander seed. *Indian Phytopath.*, 33: 478–479.

Prasad, B.K., 1980c. Incidence of preharvest endophytic seed mycoflora of some umbelliferous spices. *Indian Phytopath.*, 33: 479–480.

Prasad, T., Bilgrami, K.S., Thakur, M.K. and Singh, A., 1984. Aflatoxin problem in some common spices. *J. Indian Bot. Soc.*, 63: 171–173.

Rai, B. and Singh, D.B., 1976. Seed mycoflora of barley and its control by means of fungicides. *Proc. Indian Nat. Sci. Acad.*, 42B: 311–317.

Rai, B. and Manjari, Kanak, 1984. Efficacy of certain fungicides on seed mycoflora and seed germination of fennel (*Foeniculum vulgare* Mill). *Abst 7th All India Bot. Conference*, p. 22.

Rai, B. and Manjari, Kanak, 1985a. Studies on seed mycoflora of some. *Indian Spices Proc. 72nd Indian Sci. Congress*, p. 87.

Rai, B. and Manjari, Kanak, 1985b. Antifungal properties of extracts of some plant parts/products on stored seeds of some spices. *Abst. 8th All India Bot. Conference*, p. 34.

Rai, B. and Manjari, Kanak, 1985c. Inhibitory effect of some fungal metabolities on seed germination of fennel and ammi. *Seminar on Recent Advances in Plant Pathological Research*, Instt. of Agri. Sciences, BHU, Varanasi, p. 62.

Rai, B. and Manjari, Kanak, 1986 Effect of antibiotics on seed mycoflora and seed gemination of some Indian spices. *Proc. 73rd Indian Sci. Congress*, p. 26–27.

Randhawa, H.S. and Aulakh, K.S., 1984. Mycoflora associated with discoloured and shrivelled seeds of pearl millet. *Indian Pytopath.*, 31(1): 119–122.

Richardson, M.J., 1979. An annotated list of seed borne disease. *Phytopath* Pap. No. 23, 3rd Edn. Commonwealth Mycological Institute, Kew, England.

Scott, P.M. and Kenndy, B.P.C., 1973. Analysis and survey of ground black white and *Capsicum* peppers for aflatoxins. *J. Assoc. of Anal. Chem.*, 56: 1452–1457.

Seenappa, M., 1970. Aflatoxin *Aspergillus* and bacterial contamination in India spices involved in international trade. *Ph.D. Thesis*, University of Waterloo, Ont, Canada.

Shah, Rakesh and Jain, J.P., 1993. Seed mycoflora of mustard and its control. *Indian Journal of Mycology and Plant Pathology*, 23(3): 2991–2995.

Sheik, R.A. and Agnihotri, J.P., 1977. Anitfungal activity of some plant extract Indian. *J. Mycol. and Pl. Pathol.*, 7: 180–181.

Shrotri, S.C., Gupta, J.S. and Srivastava, R.N. Seed-borne fungi of Aster (*Callistephus chinensis* Nees) their significance and control. *J. Indian Hot. Soc.*, 65: 446–449.

Singh, Anjana, 1984. Mycotoxin problem in dry fruits and juices. In: *Abst. 7th All India Bot. Conference*, p. 27.

Singh, V., Pandey, P.C. and Jain, D.K., 1994. Spices and condiments. In: *A Textbook of Angiosperm*, pp. 110–134.

Srivastava, R.K. and Chandra, S., 1985. Studies on seed mycoflora of some spices in India–Part 3: Mycoflora of quiescent seeds spermoplane and spermoshere in relation to pre- and post-emergence mortality. In: *Proc. 72nd Indian Sci. Congress.*

Srivastava, R.N., 1984. Seed-borne fungi from Indian seed lots of *Corianders sativum* Linn. their significance and control. *J. Indian Bot. Soc.,* 63: 181–185.

Swarup, J., 1970. Effect of fungicidal treatment on the storage of seed of some unbelliferous spices. *Indian Phytopath.,* 23: 47–50.

Swarup, J. and Mathur, R.S., 1972. Seed microflora of some umbelliferous spices. *Indian Phytopath.,* 25: 125.

Swarup, J. and Tondon, I.N. Effect of microorganisms on the germination of seed of umbelliferous spices stored at different temperatures and relative humidities. *Indian Phytopath.,* 24: 615–616.

Tandon, M.P., Dwivedi, D.K. and Shukla, D.N., 1985. Effect of fungal toxins on seed germination of certain spices. In: *Proc. 72nd Indian Sci. Congress,* p.106.

Tewari, S.N. and Datt, A. Premalatha, 1984. Effect of leaf extract media of some plants on the growth of three fungal pathogens of rice. *Indian Phytopath.,* 37(3): 458–461.

Thind, T.S., Vyas, K.M. and Prakash, V., Effect of some antibiotics on the germination of coriander seeds. *Indian J. Exp. Biol.,* 15(3): 247.

Upadhyay, R.K., 1978. Biology of microfungi associated with a medicinal plant. *Ph.D. Thesis,* Banaras Hindu University, India.

Weidenborner, M., Hindrof, H. and Jha, H.C., 1990. *Aspergillus* legumes with flavonoids isoflavonoids. *Angrew Bot.,* 64: 175–190.

Chapter 13

Infliction of Seedling Diseases and Biochemical Disorders due to Storage Fungi of the Crop Seeds

B.K. Prasad[1], Anand Kishor[3], Sheo Prasad Singh[2] and Manoj Kumar[1]

[1]*University Department of Botany, Magadh University, Bodh Gaya – 824 234*
[2]*University Department of Botany, V.K.S. University, Ara – 802 301*
[3]*Department of Botany, College of Commerce, Patna – 800 020*

ABSTRACT

Of the seedlings diseases due to storage fungi of the seeds root and foot rotting, water soaked lesions in the foot, brown lesions in the root, stunting and drying of the apical buds are common in the vegetables. Failure of expansion of cotyledonary leaves and suppression of the development of chlorophyll pigment are distinct symptoms besides the above noted ones in mustard and other crucifers. Brown staining of the cotyledons with mycelial growth, puckering of the first leaf and its small size and drying of the apical bud have been reported to be the important symptoms of diseases in bean seedlings. In paddy var. *Akhanphou* (Manipur) radicles and plumule were observed to be blackish brown, basal portion of leaves with brown spots, tip of the radicle blackish and necrotic, drying of leaves and stunting were noteworthy symptoms. The number and the length of fibrous roots were less. All these symptoms were incited by species of *Aspergillus, Alternaria, Curvularia, Memnoniella, Fusarium* and other storage fungi. Among physiological and biochemical disorders in the seedlings diminution in chlorophyll

pigment, total soluble sugar and free amino acid and cation were observed besides
excited exudation of cations from the root, stimulated IAA oxidase and respiratory
enzymes, amino acid oxidase, decarboxylase and deaminase, and sluggish activity
of urease and nitrate reductase.

Introduction

Storage fungi, unequivocally inflict deterioration of the seeds resulting in their
decay and suppressing their germination. Authors have worked on different aspects
of mycodeterioration of the crop and other seeds (Christensen, 1957,1972; Anderson
et al., 1970, Anderson, 1970a).

The present epitomised review centres around the diseases associated with the
seedlings and their biochemical disorders raised from the seeds deteriorating due to
storage fungi.

Common Storage Fungi of the Crop Seeds

Dominance of *Aspergillus flavus, A. niger, A. sydowi* and *Fusarium moniliforme*
besides without considerably frequency of *A. nidulans, A. candictus, Alternaria alternata,
A. tenuissima, Curvularia lunata, C. pallescens, Cladosporium* spp. have been reported in
seeds of wheat, maize, finger millet, jowar, bajra, gram and many pulses, mustard
and other brassicaceous plants, sesame, groundnut and others. *Penicillium* spp. are
dominant on temperate regions of the world.

Effect of Storage Conditions of the Seedlot

The crop seed is stored in air tight polyethylene bags, linen bags, earthen built
storage, the "kothi" (kachcha) or cemented kothi (Pucca), conventionally, seedlots
packaged in gunny bags are stored under the heap of broken straw of crop plants
generally used as fodder. Such seedlots if properly dried in the sun in the month of
March to May of the year, is definitely safe under storage. Storage bags are also made
in remote villages by the straw of paddy fully plastered with cowdung and dried.
This sort of bag is not air tight but also not so porous to permit the entry of humid air
to the stored seedlot. The important matter is that the seedlot must be fully dried
before packaging. The seed moisture above 10 per cent level as a result of insufficient
drying provides ample opportunity of attack by fungi and their colonization on the
seeds.

Diseases in the Seedlings

A voluminous account of seedborne diseases in standing crop is available right
from the mid eighteenth century to-date (Tull, 1733, Neergaard, 1977, Richardson,
1990, Agarwal and Sinclair, 1997). The communication regarding the infliction of
diseases in the seedlings stage of crop plants and physiological and biochemical
disorders in them due to storage fungi is quite insufficient indeed (Mathur and Sehgal,
1964, Sao *et al.*, 1989, Yadav and Prasad, 1997) but recent investigations indicate that
the fungi of this nature besides deteriorating the seeds in many ways also incite
diseases in the seedlings and biochemical anomaly in them. The latter aspect has not

been appreciably worked out except the data available to-date provide sufficient ground to admit that the biochemistry of the seedlings is infact, utterly disarrayed.

The symptoms of disease in the seedlings have been observed in the root region, cotyledons and the cotyledonary and/or first leaf. Prasad (1984) in lablab bean and Kumar (2001) in some papilionaceous crops reported brown/black necrotic spots in the cotyledons of the seedlings due to *A.flavus, A. niger* and other storage fungi. Puckering of first leaf and its smaller size were found to be common in 28 per cent of the seedlings. No notable symptoms were observed in the seedlings of cucurbits except yellowing of cotyledonary leaves as keenly observed in pumpkin (Ranjan, 1995). The first leaf of bittergourd was found to exhibit puckering. This was also recorded in lady's finger besides drying of the apical bud of 12 per cent of seedlings mainly due to *A. flavus, A. niger, Alternaria tenuissima* and *A. alternata, Curvularia lunata, C. pallescens, Cladosporium oxysporum* and *Fusarium moniliforme.* The apical bud of 7-12 per cent of seedlings of tomato, brinjal and spinach dried due to these fungi. Cotyledonary leaves of these crops were smaller and somewhat pale.

The symptoms of diseases in the seedlings of crucifers due to storage fungi have been reported to be very distinct. Singh (1988) has reported partly and complete failure of expansion of cotyledonary leaves and their yellowing, smaller seedlings and smaller primary roots without branches mainly due to *Aspergillus flavus, A. niger, Fusarium moniliforme* and other storage fungi. Such symptoms were found to be more distinct in radish seedlings besides damping-off like symptoms due to shrivelling of the radicle (Sao *et al.,* 1989). Root and foot rotting were also reported in 5-9 per cent of the seedlings.

Cereals and Millets have also been worked out in the light of infliction of seedling diseases. Details have been reported in Akhanphou variety of rice (Manipur state) (Diwakar and Prasad, 1997). Yellowing and drying of the younger leaves, drying of lower leaves from apex to base and stunting were reported in aerial parts due to *Aspergillus ustus, Curvularia lunata, C. oryzae* and *Memnoniella echinata.* Percent seedling affected was more that has been raised from the seeds stored at 90 per cent RH. Radicle showed browning, necrotic and blackishess. Both the radicle and plumule were brown. Seeds stored with fungi at 80 and 90 per cent RH produced 15 to 35 per cent diseased seedlings. The authors tested the pathogenicity in vitro soaking the seeds in the metabolite of the fungi prepared in Richard solution and resultantly they found distinct suppression of growth of the radicle.

Slow rate of growth appears to be a common feature of those seedlings raised from the fungus stored seeds (Prasad and Prasad, 1989).

Biochemical Disorders in the Seedlings

Biochemical and physiological disorders have been worked out in the seedlings of the crop plants named earlier. The radicle after emergence from the seed was considered to be biochemically sick due to excited exudation of the cations such as $Ca^{++}, Zn^{++}, Fe^{++}, Co^{++}$ and Ni^{++} hexose and pentose sugars and amino acids (Kishor *et al.,* 1990). Recently Ca_4^{++}, Mg^{++} and Zn^{++} have been found to be far less in maize seedlings due to seedborne *A. flavus, A. niger* and *F. moniliforme* (Hussain, 2002). The same author reported diminutive amount of photosynthetic pigments, total soluble sugar and free amino acid in gram, maize and mustard due to the noted fungi. The author also reported smaller size of the nuclei in the cells of radicle tip of gram due to *A. flavus* and *F. moniliforme* along with significant lowering of the mitotic index.

The seedlings of mustard raised from *A. flavus* stored seeds were found to respire with rapid rate (Kishor and Prasad, 1989, Srivastava *et al.*, 1994). Sao *et al.* (1989) reported enhanced O_2 uptake and excited activity of starch phosphorylase and dehydrogenase of pyruvic acid, alpha-keto-glutaric acid and succinic acid in radish seedlings cultured in Hoagland and Knop solution mixed with the extract of those seedlings whose cotyledonary leaves failed to normaly grow and became chlorotic due to seedborne *Memnoniella echinata*.

Besides above noted changes the activity of nitrate reductase and urease that provide NH_3 from nitrate and urea respectively for the synthesis of amino acid by uniting with keto acids, has been reported to be sluggish by Yadav and Prasad (1997) in wheat and cucumber (Yadav and Prasad, 1995). The protein content of wheat grain, as a result, was found to be depleted. Similar fate of the enzymes was reported in the seedlings of mustard (Singh, 1988) and rajma (Dayal *et al.*, 2001).

The symptoms such as stunting, failure of expansion of cotyledonary leaves, drying of the apical bud and leaves, puckering, reduced growth of the seedling as a whole appear due to the toxic principles secreted by the storage fungi (Harman and Nash, 1972, Sao *et al.*, 1989). Biochemical anomalies might result due to activation of oxidative enzymes, degradation of amino acid and hindrance in their synthesis, disturbed synthesis and rapid degradation of chlorophyll, damage of plasmamembrane and obstruction of ongoing biochemical steps (Anderson, 1970a, Sao *et al.*, 1989).

The damping-off like symptom, foot and root rotting, water soaked and brown lesions in the root and staining of the cotyledons appear due to colonization of the storage with seed and seedlings possessing high pectinolytic and cellulolytic enzyme activity (Prasad, 1979).

References

Agarwal, V.K. and Sinclair, J.B., 1997. *Principles of Seed Pathology*, 2nd edn. CRC Lewis, London.

Anderson, J.D., Baker, J.E. and Warthington, E.K., 1970. Ultrastructural changes in embryos in wheat infected with storage fungi. *Plant Physiol.*, 46: 857–859.

Anderson, J.D., 1970a. Physiological and biochemical differences in deteriorating barley seed. *Crop Sci.*, 10: 36.

Christensen, C.M., 1957. Deterioration of stored grains by fungi. *Bot. Rev.*, 23: 108–134.

Christensen, C.M., 1972. Moisture content of sunflower seed in relation to invasion by storage fungi. *Plant Dis. Reptr.*, 56: 173–175.

Diwakar, A.P. and Prasad, B.K., 1997. Effect of storage fungi of seeds on the seedling diseases of Paddy. *J. Mycol. Plant Pathol.*, 27: 184–187.

Dayal, Shambhu, Kumar, Sanjay, Kumar, Manoj, Singh, S.P., Kumar, Vijendra and Prasad, B.K., 2001. Effect of storage fungi on nitrate reductase and urease activities in rajma seedlings. *Indian Phytopath.*, 54: 372–375.

Harman, G.E. and Nash, G., 1972. Deterioration of stored pea seed by *Aspergillus ruber*: Evidence for involvement of a toxin. *Phytopathology*, 62: 209–212.

Hussain, Md. Azhar, 2002. Studies on the significance of storage fungi of crop plants inflicting diseases and disorders in the seedlings. *Doctoral Thesis*, Magadh University, Bodh Gaya, (Unpublished).

Kumar, Vijendra, 2002. Studies on the seedborne fungi of papilionaceous crop and their significance. *Doctoral Thesis*, Magadh University, Bodh Gaya.

Kishor, A. and Prasad, B.K., 1990. Chlorophyll content, biomass and O_2 uptake of mustard seedlings due to storage of seeds. *J. Ind. Bot. Soc.*, 68: 407–408.

Kishor, A., Singh, R.N., Sao, R.N., Sinha, N.P. and Prasad, B.K., 1990. Physico-chemical characteristics of the root exudate of mustard seedlings raised from the seeds stored with *Aspergillus flavus*. *Indian Phytopath.*, 34: 513–516.

Mathur, R.L. and Sehgal, S.P., 1964. Fungal flora of seeds of jowar (Sorghum vulgare), its role in reduced emergence and vigour of seedling and control. *Indian Phytopath.*, 17: 227–233.

Neergaard, P., 1977. *Seed Pathology*, Vols. 1 and 2. Macmillan, London.

Prasad, A., 1984. Studies on the mycodeterioration of bean (Dolichos lab lab) seed during storage. *Doctoral Thesis*, Magadh University, Bodh Gaya.

Prasad, B.K. and Shanker, U., 1988. Effect of seedborne fungi on the growth and biochemical constituents of finger millet seedlings. *J. Mycol. Pl. Pathol.*, 18: 31 – 34.

Prasad, A. and Prasad B.K., 1989. Effect of toxic substance extracted from *A. niger* inoculated seeds on the growth of radicle and plumule of Dolichos lab lab. *J. Indian Bot. Soc.*, 68: 107–108.

Prasad, B.K., 1979. Enzymic studies of seedborne fungi of coriander. *Indian Phytopath.*, 32: 92–94.

Ranjan, Sudhir, 1995. Studies on the physiological disorder and diseases of seedlings of some vegetables caused by the seedborne storage fungi. *Doctoral Thesis*, Magadh University, Bodh Gaya.

Richardson, M.J., 1990. *An Annotated List of Seedborne Disease*. ISTA, Secretariat, Zurich, Switzerland.

Singh, Sheo Prasad, 1989. Studies on the seedling diseases of mustard due to storage mould. *Doctoral Thesis*, Magadh University, Bodh Gaya.

Srivastava, A.K., Naresh, Ram, Kumar, Sanjay and Prasad, B.K., 1994. Change in the respiration of mustard seedlings due to storage of seed with *Aspergillus* flavus. *Bulletin of Pure and Applied Sciences*, 13: 15–19.

Tull, J., 1733. The Horse Hoing Husbandry: Or an essay on the principles of fill age and vegetation. Chapter XII of Smutiness. G. Strabon. Cornhill, England.

Yadav, B.N. and Prasad, B.K., 1997. Nitrate reductase and urease activity in wheat due to faulty storage of seed. *J. Indian Bot. Soc.*, 76: 59–61.

Yadav, B.N. and Prasad, B.K., 1995. Urease and nitrate activity in cucumber seedlings due to faulty storage of seed. *Columbia J. of Life Sciences*, 5: 44–46.

Chapter 14

Antimicrobial Potentials of some Important Spices Plants of India

Leena Parihar* and A. Bohra

Microbiology laboratory, Department of Botany,
J.N.V. University, Jodhpur – 342 001

India is profusely rich in the history of medicinal plants and its 75 per cent folk population is still using herbal preparations in the form of powder, extracts and decoction because these are easily available in nature and the natives have stronger faith on traditional plants. Ministry of Health and Family Welfare Center and state governments is conducting high-level research programmes to manufacture drugs. These drugs of medicinal value are competing today in markets. Various plants exhibit various types of antimicrobial activities. Even parasitic plants and Orchids also are of great medicinal value, which are found to be antimicrobial.

Spices by virtue of their possessing great variety and fascinating foliage have drawn the attention and admiration of agriculturists for centuries. Medicinal value of these plants is known to man for more than 2000 years. Various spices plant parts have been mentioned in Sushurta and Charak Samhita. Besides these various researchers also found out and screened various spice plants for their antimicrobial activities.

Herbs and spices have been valued for ages not only for their culinary uses but also for their medicinal properties. These provided the materials used in various systems of alternate medicine such as Ayurveda, Siddha, Unani, Chinese, Tibetan,

---

* Corresponding Author E-mail: pradeepparihar2002@yahoo.com.

Naturopathy, Aromapathy, Homoeopathy and Flower remedies. Ayurveda developed to its peak form some two thousand years ago and widely practiced for centuries thereafter fell to the modern antibiotics and synthetic generic drugs in the twentieth century. Indeed, traditional knowledge of folk and tribal medicine has gifted several modern drugs to civilized man, as best exemplified by quinine from Cinchona bark and Salycylate (the mother compound for aspirin) from a willow tree.

Spices have been valued for culinary uses but they are also used in Medicinal properties. Out of 3,00,000 species of higher plants available. Only a small portion has been investigated for medicinal properties and still smaller number yield well defined drugs. The medicinal value of spices is not so great as was though during the middle ages, but a considerable number of them are still official drugs in both Europe and America. They are used as Carminatives and antiseptics.

The same arguments apply to many spices, which have medicinal values. The uses of turmeric or chili, pepper or garlic are commonly known. What does general public not know is that even the commonly prevailing notion of chili being called 'hot' has not been shown to have a profound scientific basis. Capscicin (the hot principle of chili) activates that promoter of a brain receptor gene, which is also activated by heat. Curcumin (the active principle of Haldi, *Curcuma longa*) is not only an antibiotic, for which it is used in boils or ulcers, but also has cancer retarding or prevailing properties. Many such examples can be cited for the existence of curing principles in spices.

Chemically, depending on their active principles, plants may have alkaloids, glycosides, steroids or other groups of compounds, which may have marked pharmaceutical actions as anticancerous, etc. Many of the essential oils, dyes, lattices and even tannins and vegetable oils are also widely used as medicines. The many substances that go into making up medicine are frequently products of living cells, although seemingly 'waste' or intermediate, metabolic compounds and not an integral part of the protoplasm and may have no obvious utility to the plants.

Out of nearly 3,00,000 species of higher plants available, only a small proportion have been investigated for medicinal properties, and a still smaller number yield well-defined drugs. The same is the case with lower plants and with plants of the sea. Thus, the knowledge of plant constituents gained so far is still meager, considering the huge number of species available in the world. Approximately, only 10 per cent of the organic constituents of plants are reported to be known and the remaining 90 per cent are yet to be explored.

Very small proportions of Indian medicinal plants are lower plants like lichens, ferns, algae, etc. The majority of medicinal plants are higher plants. The major families in which medicinal plants occur are Fabaceae, Euphorbiaceae, Asteraceae, Poceae, Rubiaceae, Cucurbitaceae, Apiaceae, Convolvulaceae, Malvaceae and Solanaceae.

Shekhawat and Prasada (1971) described the antifungal properties of some spices plant extracts and also found out the inhibition of spore germination by these extracts. Prasad *et al.* (1972) studied the *in vitro* antimicrobial activities of leaf oil of *Piper betle* and found it inhibitory against *E. coli, B. proteus* and *S. pneumoniae.* Kotte and Shinde (1973) have studied the influence of some spices plant extracts of certain hosts on the growth and sclerotial formation of *Macrophomina phaseolina in vitro*, Agarwal *et al.* (1979) studied the antifungal activity of some *Terpenoids* against the effect of some plant extracts on the conidial germination of *Curvularia pallescens*.

Gupta and Singh (1983) studied the effect of certain plant extracts and chemicals on teliospore germination of *Neovossia indica*. Kishore *et al.* (1983) studied the volatile fungi toxicity in some higher plants as evaluated against *Rhizactonia solani* and some other fungi. Singh *et al.* (1983) studied the fungitoxic properties of essential oil of *Mentha arvensis*. Sharma *et al.* (1984) studied the microbiological status and antifungal properties of some irradiated spices. Shamshad *et al.* (1985) have done microbiological studies on some commonly used spices in Pakistan.

Karapinar (1986) studied the effects of Citrus oils and some spices on growth and aflatoxin production by *Aspergillus parasiticus*. Kivanc and Akgul (1986) described the antibacterial activities of essential oils from Turkish spices and Citrus. They found these extracts inhibiting the growth of nearly seven bacteria including *Staphylococcus aureus* and *Proteus vulgaris*. Dubey and Tripathi (1987) studied the antifungal, physio-chemical and phytotoxic properties of essential oils of *Piper betle*. Prasad (1987) studied the effect of seed extract of *Coriander* inoculated with *Aspergillus flavus* on the hydrolytic enzymes of food resources.

Shukla and Tripathi (1987) studied the physio-chemical, phytotoxic and fungitoxic properties of essential oil of *Foeniculum vulgare*. Jagannathan and Narasimhan (1988) studied the effect of spices plant extracts or their products on two fungal pathogens of finger millet. Zaika (1988) determined the antimicrobial activities of various spices and herbs. Akgul and Kivanc (1989) studied the antibacterial activity of spices, sorbic acid and sodium chloride against seven foodspoilage bacteria. Lakshmanan (1990) studied the effect of certain spices plant extracts against *Corrynespora cassiicola*. Upadhyaya *et al.* (1990) studied the effect of extracts of some medicinal plants including few spices plants on the growth of *Curvularia lunata*.

Chauhan and Singh (1991) studied the effect of volatiles of some plant extracts on germination of zoospores of *Phytophthora drechsleri*. Dube *et al.* (1991) studied the fungitoxic and insect repellent efficacy of some spices against *Aspergillus* sps. Liewen *et al.* (1991) described some antifungal food additives including some spices. Salmeron and Pozo (1991) described the effect of Cinnamom *(Cinnamomum zeylanicum)* and clove *(Eugenia caryophyllus)* on growth and toxigenesis of *Aspergillus flavus*. Singh *et al.* (1991) described the controlling measures of powdery mildew of pea by ginger extract.

Garg and Jain (1992) and Garg and Siddiqui (1992) described the biological activity of essential oils of *Piper betle* and *Coriandrum sativum* and found it toxigenic against various plants as well as human pathogenic bacteria and fungi. Rana and Joshi (1992) investigated the antiviral activity of ethanolic extracts of *Syzygium* sps. Adegoke and Sagma (1993) studied the influence of different spices on the microbial reduction and storability of laboratory processed tomato ketch up and menced meat. Baby *et al.* (1993) studied the *in vitro* antimicrobial activities of leaf oil of *Piper betle* against pathogenic microorganisms in pure form and at various dilutions.

Ishrat *et al.* (1994) studied the effect of turmeric derivatives on radial colony growth of different fungi. Meera and Sethi (1994) studied the antimicrobial activity of essential oils from spices against various microorganisms including human pathogens. Mohammed *et al.* (1994) studied the antimicrobial activity of *Piper betle* and other Malaysian plants against fruits pathogens and suggested various compounds to be present in those parts of plants. Shetty *et al.* (1994) studied antimicrobial activity of Cumin against *Aspergillus sps., Saccharomyces Cerevisiae* and aflatoxin on storage corn seeds by medicinal and spices extracts. Singh *et al.* (1994) studied the antifungal activity of *Mentha spicata* against *Fusarium oxysporum.* Tiwari *et al.* (1994) studied the fungitoxic properties of essential oil of *Cinnamomum zeylanicum* against *Aspergillus* sps. Yadav and Dubey (1994) screened some essential oils of spice plants against ringworm fungi, *Trichophyton mantagrophytes.*

Apisariyakul *et al.* (1995) studied the antifungal activity of turmeric oil extracted from *Curcurma longa* against various dermatophytes like *Trichophyton* etc. Bara and Vanetti (1995) studied the antimicrobial activity of *Curcuma longa* leaves against gram negative bacteria and pathogenic fungi.

Bhatti (1996) studied the antimicrobial activity of *Trigonella foenum-graceum* seeds. Mukherjee *et al.* (1996) studied the antifungal activities of leaf extracts of *Cassia tora.* Tsakala *et al.* (1996) screened *in vitro* antibacterial activity from *Syzygium guinesense* hydrosoluble dry extract against *Salmonella, Shigella, Entrobacter* etc. Tiwari (1997) studied the fungitoxicity of volatile constituents of some higher plant against pre-dominant storage fungi. Purohit and Bohra (1996) studied effect of some spices plant extract on conidial germination of some important phytopathogenic fungi. Mohta *et al.* (1998) studied the antimicrobial activity of 38 species of Malaysian plants including spices plants against various dermatophytes. Nidiry (1998) studied the structure fungitoxity relationship of the monoterpenoids of the essential oils of Peppermint and scented geranium. Nalini *et al.* (1998) studied the influence of spices on colon cancer. Rajendharan *et al.* (1998) studied the antimicrobial activity of some spices and medicinal plants against *E. coli, Staphylococcus aureus, Pseudomonus aeroginosa* and *Klebsiella* sps. Saju *et al.* (1998) studied the antifungal and insect repellent property of essential oil of *Curcuma longa.* Purohit and Bohra (1999) studied antifungal activity of some spices plants against *Fusarium oxysporum.*

Arora *et al.* (1999) and Bedin *et al.* (1999) described the antimicrobial activities of various spices plants. Brindani and Bacci (1999) studied the inhibitory action of spices on *Yersinia entrocolitica* and Bacillus *circus* isolated from foods of animal origin. Lean and Mohammed (1999) studied the antimycotic effects of turmeric, clove, black pepper leaves etc. Mina *et al.* (1999) studied the inhibitory effect of *Cinnamom* extracts on *Helicobacter pylori.* Radha *et al.* (1999) studied the antifungal properties of crude leaf extracts of *Syzygium.* Sharma and Nanda (2000) studied the effect of various plant extracts including some spices plants on teliospore germination of *Neovossia indica.* Vijayaraghhavan and Rajkumar (2000) studied antimicrobial activity of herbal mixture over *Staphylococcus aureus.* Kaushik and Yograj Singh (2000) have explained about the antibacterial activity of extract of rhizome of *Curcuma longa* against bacteria! strain of *Escherichia coli.*

To find out the mechanism of Pharmacognosy of various diseases, researchers have been found out the phytochemistry of some spices but very less work has been done on these spices.

Ramchandraiah *et al.* (1986) studied the essential and fatty oil content in Umbelliferous and Fenugreek seeds and found out the percentage of essential oils present in them. Venkatenarayana *et al.* (1989) studied the formulation and evaluation of herbal vanishing cream. They evaluate *Curcuma longa* for antibacterial activity and its phytochemistry. Jain *et al.* (1992) studied the pharmacological evaluation of *Cuminum cyminum*. Gowri and Regupathy (1994) studied the multi residue analysis of insecticides in spices by GC–ECD. In this chilli, ginger and turmeric have been studied. Dung *et al.* (1995) have done chemical investigation of the aerial parts of *Zingiber zerumbet* and found various types of essential oils present in it. Gupta *et al.* (1996) analysed structural carbohydrates and mineral contents of fenugreek seeds. Chung *et al.* (1997) studied the hydroxyl radical scavenging effects of spices.

Till today, very less efforts have been made to test *in vitro* antimicrobial activity of spice plant as well as analysis of their active phytoconstituents. In the present investigation an attempt has been made to test the antimicrobial activity of some important spices plants against human pathogenic organisms.

From all these above observation it has been found that Spices plants constitutes various antimicrobial activities against various types of microorganism. These observations suggests that these phytochemical constituents including alkaloids, phenols, tannins, sterols etc. are known for their antimicrobial actions. Thus the combined effect against of these phytoconstituents may be responsible for the synergistic antimicrobial activity of these plant extracts. Further activity guided fractionation of crude extracts and their interactions with different active fractions of the plants are needed to explore the exact mechanism of interaction among the active phytoconstituents. Similarly the efficacy of crude extracts or polyherbal preparation needs to be studied *in vivo* assess their therapeutic utility.

References

Adegoke, G.O., Sagua, V.Y., 1993. Influence of different spices on the microbial reduction and storability of laboratory processed tomato ketchup and minced meal. *Nahrung,* 37(4): 352–355.

Agarwal, I., Mathela, C.S. and Sinha, S., 1979. Studies on the antifungal activity of terpenoids against *Aspergilli. Indian Phytopathology,* 32(1): 104–105.

Agarwal, Sanjeev and Agarwal, S., 1996. Volatile oil constituents and wilt resistance in Cumin (*Cuminum cyminum* L.). *Current Science,* 71(3): 177–178.

Akgul, A. and Kivanc, M., 1989. The antibacterial effect of spices, sorbic acid and sodium chloride. Doga, Turk, tarim ve, Ormancilik, Dergisi, 13(1): 1–10.

Akhtar, M.A., Rahber, Bhatti, M.H. and Aslam, M., 1997. Antibacterial activity of plant diffusate against *Xanthomonas campestris* pv. *Citri. International Journal of Pest Management,* 43(2): 149–153.

Apisariyakul, A., Vanittanakom, N. and Buddhasukh, D., 1995. Antifungal activity of turmeric oil extracted from *Curcuma longa*. *Journal of Ethnopharmacology*, 49(3): 163–169.

Arias, M.L., Vtzinger, D. and Monge, R., 1997. Microbiological quality of some powder spices of common use in Costa Rica. *Revista-de-Biologia-Tropical*, 44–45: 3–11.

Arora, D.S., Kaur, Jasleen and Kaur, J., 1999. Antimicrobial activity of spices. *International Journal of Antimicrobial Agents*, 12(3): 257–262.

Baby, P., Alankararao, G.S.J.G. and Prasad, Y.R., 1993. Leaf oil of *Piper betle* Linn. the *in vitro* antimicrobial studies. *Indian Perfumer*, 37(1): 91–93.

Bara, M.T.F. and Vanetti, M.C.D., 1997. Antimicrobial effect of spices on the growth of *Yersinia enterocolitica*. *Journal of Herbs, Spices and Medicinal Plants*, 3(4): 51–58.

Bedin, C., Gutkoshi, S.B. and Wiest, J.M., 1999. Antimicrobial activity of spices. *Higiene-Alimentar*, 13(65): 26–29.

Bhatti, M.A., Khan, M.T.J., Ahmed, B., Jamshaid, M. and Ahmed, W., 1996. Antibacterial activity of *Trigonella foenum-graecum* seeds. *Fitoterapia*, V. 67(4): 372–374.

Biling, J. and Sherman, P.W., 1998. Antimicrobial functions of spices: why some like it hot. *Quarterly Review of Biology*, 73(1): 3–49.

Brindani, F. and Bacci, C., 1999. Inhibitory action of spices on *Yersinia enterocolitica* and *Bacillus circus* isolated from foods of animal origin. *Annali-della-Facolta-di-Medicina-Veterinaria-Universita-di-Parma*, 19: 297–306.

Cai, L. and Wu, C.D., 1996. Compounds from *Syzygium aromaticum* possessing growth inhibitory activity against oral pathogens. *Journal of Natural Products*, 59(9): 987–990.

Chauhan, V.B. and Singh, U.P., 1991. Effect of volatiles of some plant extracts on germination of zoospores of *Phytophthora drechsleri* f. sp. *Cajani*. *Indian Phytopathology*, 44(2): 197–200.

Chung, S.K., Osawa, T. and Kaneakishi, S., 1997. Hydroxyl radical scavenging effects of spices and scavengers from brown mustard (*Brassica nigra*). *Bioscience, Biotechnology and Biochemistry*, 61(1): 118–123.

Daswani, L. and Bohra, A., 2000. Antibacterial potential of *Elettaria Cardamomum* (Chotti Elaichi) against human pathogenic bacteria, *Staphylococcus aureus*. *Journal of Eco-physiology*, 3(3–4): 135–137.

Daswani, L. and Bohra, A., 2002. Antibacterial efficacy of extracts of various spices: A study *in vitro*. *Science and Culture*, 68(1–4): 99–100.

Daswani, L., Parihar, P. and Bohra, A. 2002. Effect of extracts of *Trigonella foenum-graecum* Linn. on the growth of *Aspergillus flavus* and *Fusarium oxysporum*. In: *Abs. Proc. of National Symposium on Arid Legumes for Food, Nutrition Security and Promotion of Trade*, May 15–16, p.193–194.

Daswani, L. and Bohra, A., 2002. Antibacterial screening of different plant part extracts of some common spices plants against human pathogenic bacteria–*Salmonella typhi.* In: *Abs. Proc. of National Seminar on Role of Antimicrobials for Sustainable Horticulture,* January 20, p. 12–13.

Daswani, L., Parihar, P., Bohra, A. and Bohra, S.P., 2002. Antibacterial activity of *Riccia aravelliensis* (Pandey et Udar) and *Plagiochasma appendiculatum* (Lahm et Lindenb). *Bio Sciences Research Bulletin,* 18(1): 61–63.

Daswani, L. and Bohra, A., 2002. Antibacterial effect of Fennel (*Foeniculum vulgare*) on the growth of human pathogenic strain of *Staphylococcus aureus. Advances in Plant Sciences,* 15(2): 579–586.

Daswani, L. and Bohra, A., 2003. Toxic effect of *Elettaria cardamomum* (Choti Elaichi) on the growth of *Salmonella typhi. Advances in Plant Sciences,* 16(1): 85–87.

Daswani, L. and Bohra, A. Antibacterial activity of various plant parts extracts of some spices plants against human pathogenic strain of *Staphylococcus aureus. Science and Culture* (In press)

Daswani, L. and Bohra, A. Antibacterial and Phytochemical Marvels of Fennel (*Foeniculum vulgare*). *Eco-research Journal of Biosciences,* 2(1) (In press).

Deans, S.G., Noble, R-C., Hiltunen, R., Wuryani, W. and Penzes, L.G., 1995. Antimicrobial and antioxidant properties of *Syzygium aromaticum* (L). Merr. and Perry. Impact upon bacteria, fungi and fatty acid levels in ageing mice. *Flavour and Fragrance Journal,* 10(5): 323–328.

Dubey, P. and Tripathi, S.C., 1987. Studies on antifungal, physio-chemical and phytotoxic properties of the essential oils of *Piper betle. Journal of Plant Dis. Prot.,* 94(3): 235–241.

Dube, S., Upadhyaya, P.D. and Tripathi, S.C., 1991. Fungitoxic and insect repellent efficacy of some spices. *Indian Phytopathology,* 44(1): 101–105.

Dung, N.X., Chinh, T.D. and Leclercq, P.A., 1995. Chemical investigation of the aerial parts of *Zingiber zerumbet* (L) Sm. from Vietnam. *Journal of Essential Oil Research,* 7(2): 153–157.

Garg, S.C. and Jain, R., 1991. The essential oil of *Zingiber officinale* Rose: A potential insect repellent. *Journal of Economic Botany and Phytochemistry,* 2(1–4): 25–27.

Garg, S.C. and Jain, R., 1992. Biological activity of the essential oil of *Piper betle. Journal of Essential Oil Research,* 4(6): 601–606.

Garg, S.C. and Siddiqui, N., 1992. *In vitro* antifungal activities of the essential oil of *Coriandrum sativum* Linn. *Journal of Research and Education in Indian Medicine,* 11(3): 11–13.

Gowri, T. and Regupathy, A., 1994. Multiresidue analysis of insecticides in spices by GC–ECD. *PLACROSYM,* Calicut, 11: 88.

Gupta, Kaushalya, Thakral, K.K., Arora, S.K., Chowdhary, M.L. and Gupta, K., 1996. Structural carbohydrate and mineral contents of Fenugreek seeds India-cocoa. *Arecanut and Spices Journal,* 20(4): 120–124.

Gupta, R.P. and Singh, A., 1983. Effect of certain plant extracts and chemicals on teliospore germination of *Neovossia Indica*. *Indian Journal of Mycology and Plant Pathology*, 13(1): 116–117.

Ishrat, Niaz, Kazmi, A.R. and Jilani, Ghulaam, 1994. Effect of turmeric derivatives on radical colony growth of different fungi. *Sarhad J. of Agriculture*, 10(5): 571–573.

Iyenger, M.A., Rama Rao, M., Bairy, I. and Kamath, M.S., 1995. Antimicrobial activity of the essential oil of *Curcuma longa* leaves. *Indian Drugs*, 32(6): 249–250.

Jagannathan, R. and Narasimhan, V., 1988. Effect of Plant Extracts/Products on two fungal pathogens of finger millet. *Indian Journal of Mycology and Plant Pathology*, 18(3): 250–254.

Jain, S.C., Purohit, M. and Jain, R., 1992. Pharmacological evaluation of *Cuminum cyminum*. *Fitoterapia*, 63(4): 291–294.

Kapoor, A., 1997. Antifungal activities of fresh juice and aqueous extracts of turmeric (*Curcuma longa*) and ginger (*Zingiber officinale*). *J. of phytol. Res.*,10(1–2): 59–62.

Karapinar, M., 1996. The effect of citrus oils and some spices on growth and aflatoxin production by *Aspergillus parasiticus* NRRL 2999. *International J. of Food Microbiology*, 2(4): 239–245.

Kaushik, P. and Singh, Yograj, 2000. Antibacterial activity of extract of rhizome of *Curcuma longa* (turmeric). *J. Indian Bot. Soc.*, 79: 191–192.

Kishore, N., Dubey, N.K., Srivastava, O.P. and Singh, S.K., 1983. Volatile fungitoxicity in some higher plants as evaluated against *Rhizoctonia solani* and some other fungi. *Indian Phytopathology*, 36(4): 724–726.

Kivanc, M. and Akgul, A., 1986. Antibacterial activities of essential oils from Turkish spices and citrus. *Flavour and Fragrance Journal*, 1(4–5): 175–179.

Kotte, S.J. and Shinde, P.A., 1973. Influence of plant extracts of certain hosts on the growth and sclerotial formation of *Macrophomina phaseolina in vitro*. *Indian Phytopathology*, 26(2): 351–352.

Kumar, Rajiv and Sachan, S.N., 1979. Effect of some plant extracts on the conidial germination of *Curvularia pallescens*. *Indian Phytopathology*, 32(3): 489.

Kurucheve, V. and Padmavathi, R., 1997. Fungitoxicity of selected plant products against *Pythium aphanidermatum*. *Indian Phytopathology*, 50(4): 529–535.

Lakshmanan, P., 1990. Effect of certain plant extracts against *Corrynespora cassicola*. *Indian J. of Mycology and Plant Pathology*, 20(3): 267–269.

Lean, L.P. and Mohammed, S., 1999. Antioxidative and Antimycotii effects of turmeric, lemon-grass, betel leaves, clove, black pepper leaves and Garcinia atriviridis butter cakes. *Journal of the Science of Food and Agri.*, 79(3): 1811–1822.

Liewen, M.B., Arora, D.K., Mukerji, K.G. and Marth, E.M., 1991. Antifungal food additives. *Handbook of Applied Mycology*, 3: 541–552.

Mandal, R.K., 2001. Herbs and spices: Resources for millenium drugs. *Science and Culture*, 67: 1–2.

Meera, M.R. and Sethi, V., 1994. Antimicrobial activity of essential oils from spices. *Journal of Food Science and Technology*, 31(1): 68–70.

Mina, Tabak, Arnon, R. and Ishaak, Neeman, 1999. *Cinnamon* extracts inhibitory effect on *Helicobacter pylori*. *J. of Ethnopharmacology*, 67(3): 269–277.

Minakshi, De., De, A.K., Banerjee, A.B. and De, M., 1999. Antimicrobial screening of some Indian spices. *Phytotherapy-Research*, 13(7): 616–618.

Mohammed, S., Saka, S., El-Sharkawi, S., Ali, A.M. and Muid, S., 1994. Antimicrobial activity of *Piper betle* and other Malaysian plants against fruit pathogens. *ASOMPS, III*, Malaysia.

Mohtar, M, Shaari, K., Ali, N.A.M., and Ali, A.M., 1998. Antmicrobial activity of selected Malaysian plants against microorganisms related to skin infections. *Journal of Tropical Forest Products*, 4(2): 199–206.

Mukherjee, P.K., Saha, K, Saha, B.P., Pal, M. and Das, J., 1996. Antifungal activities of the leaf extract of *Cassia tora* Linn. *Phytotherapy Research*, 10(6): 521–522.

Muneem, K.C, Verma, S.K. and Pant, K.C., 1995. Performance of some chillies against leaf spot, powdery mildew and fruit rot in Kumaon Hills. *Indian Phytopathology*, 41(2): 206.

Nidiry, E.S.J., 1998. Structure-fungi toxicticity relationship of the monoterpenoids of the essential oils of Peppermint *(Mentha piperita)* and scented geranium *(Pelagonium graveolens)*. *Journal of Essential Oil Research*, 10(6): 628–631.

Naini, N., Sabitha, K., Vishwanathan, P. and Menon, V.P., 1998. Influence of spices on the bacterial (enzyme) activity in experimental colon cancer. *Journal of Ethnopharmacology*, 62(1): 15–24.

Pelevitch, D., Craker, L.E., 1995. Nutritionl and Medicinal importance of red pepper (*Capsicum* spp). *Journal of Herbs, Spices and Medicinal Plants*, 3(2): 55–83.

Peter, K.V., 1998. Spices Research in India. *Indian Journal of Agri. Sci.*, 68(8): 527–532.

Prasad, B.K., 1987. Effect of seed extract of Coriander inoculated with *Aspergillus flavus* on the food reserve. *Indian Phytopathology*, 40(2): 105.

Prasad, V.R., Baby, P., Alankararao, G.S., J.G., 1972. Leaf oil of *Piper betle* Linn: The *in vitro* antmicrobial studies. *Perfumerie and Kosmetic*, 73(8): 544.

Pruthi, J.S., 2001. Recent advances in spice processing technology. *Science and Culture*, 67: 10–29.

Purohit, P. and Bohra, A., 1998. Effect of some plant extract on conidial germination of some important pathogenic fungi. *Geobios News Reports*, 17: 183–184.

Purohit, P. and Bohra, A., 1999. Antifungal activity of extract from spice plants against *Fusarium oxysporum*. *Geobios News Reports*, 18: 151–152.

Radha, R., Mohan, M.S.S. and Anand, A.S., 1999. Antifungal properties of crude leaf extracts of *Syzygium travancoricum*. *Journal of Medicinal and Aromatic Plant Sciences*, 21(Suppl. 1): 55–56.

Raja, J. and Kurucheve, V., 1996. Influence of plant extracts on the growth and sclerotial germination of *Macrophomina phaseoline*. *Indian Phytopathology*, 51(1): 102–103.

Rajendhran, J., Arun Mani, M. and Navaneethakanan, K., 1998. Antmicrobial activity of some selected medicinal plants. *Geobios*, 25(4): 280–282.

Ramachandraiah, O.S., Reddy, P.N., Azeemoddin, G., Ramayya, D.A. and Rao, S.D.T., 1986. Essential and Fatty oil content in Umbelliferous and Fenugreek seeds of Andhra Pradesh habitat. *Indian Cocoa, Arecanut and Spices Journal*, 10(1): 1–12.

Rana, N.S. and Joshi, M.N., 1992. Investigation on the antiviral activity of ethanolic extracts of *Syzygium* sps. *Filoterapia*, 63(3): 542–544.

Rath, C.C., Dash, S.K., Mishra, R.K., Ramchandraiah, O.S., Azeemoddin, G. and Charyulu, J.K., 1999. A note on the characterization of susceptibility of turmeric (*Curcuma longa*) leaf oil against *Shigella* sps. *Indian Drugs*, 36(2): 133–136.

Saju, K.A., Venugopal, M.N. and Mathew, M.J., 1998. Antifungal and insect-repellent activities of essential oil of turmeric (*Curcuma longa* L). *Current Sciences*, 75(7): 660–662.

Salmeron, J. and Pozo, R., 1991. Effect of Cinnamon (*Cinnamomum zeylanicum*) and clove (*Eugenia caryophyllus*) on growth and toxigenesis of *Aspergillus flavus*. *Microbiologie, Ailments, Nutrition*, 9(1): 83–87.

Sarma, Ranjana, Phookan, A.K. and Bhagbati, K.N., 1999. Efficacy of some plant extracts in the management of shealth blight disease of rice. *Indian Journal of Mycology and Plant Pathology*, 29(3): 336–339.

Shamshad, S.K., Zuberi, R. and Qadri, R.B., 1985. Microbiological studies on some commonly used spices in Pakistan. *Pakistan Journal of Scientific and Industrial Research*, 28(6): 395–399.

Sharma, A., Ghanekar, A.S., Padwal, Desai, S.R. and Nadkarni, G.B., 1984. Microbiological status and antifungal properties of irradiated spices. *Journal of Agricultural and Food Chemistry*, 32(5): 1061–1063.

Sharma, Indu and Nanda, G.S., 2000. Effect of Plant extracts on teliospore germination of *Neovossia indica*. *Indian Phytopathology*, 53(3): 323–324.

Shekhawat, P.S. and Prasada, R., 1971. Antifungal properties of some plant extracts. I. Inhibition of spore germination. *Indian Phytopathology*, 24(4): 800–802.

Shetty, R.S., Singhal, R.S. and Kulkarni, P.R., 1994. Antimicrobial properties of Cumin. *World Journal of Microbiology and Biotechnology*, 10(2): 232–233.

Shivpuri, Asha, Sharma, O.P. and Jhamaria, S.L., 1997. Fungitoxic properties of plant extracts against pathogenic fungi. *Indian Journal of Mycology and Plant Pathology*, 27(1): 29–31.

Shukla, H.S. and Tripathi, S.C., 1987. Studies on physio-chemical, phytotoxic and fungitoxic properties of essential oil of *Foeniculum vulgare*. *Mill, Beitr. Biol. Pflanzen*, 62: 149–158.

Sinchaisri, P., Prathuangwong, S. and Eamsiri, A., 1994. Control of *Aspergillus flavus* and aflatoxin on storage corn seeds by medicinal and spices extracts. *ASOMPS*, VIII, Malaysia.

Sindhan, G.S., Hooda Indra and Parashar, R.D., 1999. Evaluation of plant extracts for the control of powdery mildew of pea. *Indian Journal of Mycology and Plant Pathology*, 29(2): 257–258.

Singh, A.K., Dikshit, A. and Dixit, S.N., 1983. Fungitoxic properties of essential oil of *Mentha arvensis* var, *Piperescens. Perfumes and Flaverist*, 8: 55–58.

Singh, J., Dubey, A.K. and Tripathi, N.N., 1994. Antifungal activity of *Mentha spicata. International Journal of Pharmacognosy*, 32(4): 314–319.

Singh, U.P., Srivastava, B.P., Singh, K.P. and Mishra, G.D., 1991. Control of powdery mildew of pea by ginger extract. *Indian Phytopathology*, 44(1): 55–59.

Sivropoulou, A., Kokkini, S., Lanaras, T. and Arsenakis, M., 1955. Antimicrobial activity of mint essential oils. *Journal of Agricultural and Food Chemistry*, 43(9): 2384–2388.

Sunderraj, T., Kurucheve, V. and Jayaraj, J., 1996. Screening of higher plants and animal faeces for the fungitoxicity against *Rhizoctonia solani. Indian Phytopathology*, 49(4): 398–403.

Tiwari, Ramesh, 1997. Fungitoxicity of volatile constituents of some higher plants against predominant storage fungi. *Indian Phytopathology*, 50(4): 548–551.

Tiwari, R., Dixit, R., Dixit, S.N., 1994. Studies on fungitoxic properties of essential oil of *Cinnamomum zeylanicum Breyn. Indian Perfumer*, 38(3): 98–104.

Tsakala, T.M., Penge, O., John, K, 1996. Screening of *in vitro* antmicrobial activity from *Syzegium guineense* (wiled) hydrosoluble dry extract. *Annales Pharmaceutiques Francaise*, 54(6): 276–279.

Umadevi, I. and Daniel, M., 1990. Phenolics of some fruit spices of the Apiaceae. *Nat. Acad. Sci. Letters.*, 13(12): 439–441.

Umadevi, I. and Daniel, M., 1993. Phenolics of some Indian Spices. *Acta Botanica Indica*, 21: 262–266.

Upadhyaya, Manju Lata and Gupta, R.C., 1990. Effect of extracts of some medicinal plants on the growth of *Curvularia lunata. Indian Journal of Mycology and Plant Pathology*, 20(2): 144–145.

Venkatanarayana, V., Kokate, C.K. and Venkateswaralu, V., 1989. Formulation and evaluation of herbal vanishing cream. *Indian Medicine*, 1(2): 6–11.

Vijayaraghvan, R. and Rajkumar, R., 2000. Antimicrobial activity of Herbal mixture over *Staphylococcus aureus. Microbiotech.*, pp. 215.

Viswanathan, K., 1995. Survey on Medicinal Spices of the Nilgiris. *Ancient Sciences of Life*,14(4): 258–267.

Yadav, P. and Dubey, N.K., 1994. Screening of some essential oils against ringworm fungi. *Indian Journal of Pharmaceutical Sciences*, 56(6): 227–230.

Zaika, L.L., 1988. Spices and herbs: The antimicrobial activity and its determination. *Journal of Food Safety*, 9(2): 97–118.

Chapter 15

A Criticism of Sareen and Wadhwa's (1981) Paper Entitled, Embryological Studies in Papilionaceae: The Genus *Alysicarpus Neck*–A Critical Review

S.A. Salgare

Department of Botany, Institute of Science, Mumbai – 400 032

ABSTRACT

In *Alysicarpus vaginalis* the male archesporium is multi-cellular and hypodermal. The anther wall is four-layered and its development confirmed to the Dicotyledonous type. The tapetum is uninucleate, uni-seriate and secretory type. The endothecium forms fibrous thickenings at maturity. Simultaneous cytokinesis results in tetrahedral and isobilateral tetrads. Pollen grains are shed at bi-nucleate and bi-celled stage. Some pollen grains show the sign of germination before anthesis and had three-nuclei. However, Sareen and Wadhwa (1981) reported uni-cellular archesporium, Monocotyledonous type of anther wall development and decussate microspore tetrads. They also failed to report three-nucleate pollen and their germination *in situ*. The ovule is bitegmic, crassinucellate and campylotropous. The female archesporium which is uni- or bi-cellular is hypodermal in origin. A linear tetrad of megaspores is formed. The development of the megagametophyte is confirmed to the Polygonum type. Some abnormalities were observed during the development of the megagametophyte, indicating that the nuclear divisions in the megagametophyte are not always simultaneous resulting in three-, five- and six-nucleate megagametophytes. At one instance in the eight-nucleate an anomalous megagametophyte, polar nuclei were missing and two extra antipodals were found arranged in two series (3+2). However, Sareen and Wadhwa (1981) were not aware of any type of anomaly in the megagametophyte. Fertilization is porogamous. Though double fertilization is a rule occasionally single fertilization that is syngamy occurred without triple fusion. Very often the zygotic nucleus was found to divide prior to the primary endosperm nucleus. The endosperm development follows the nuclear type. Sareen and Wadhwa (1981) were unaware of such anomalies. In *Alysicarpus vaginalis* six different Megarchtypes (A_1, A_2, B_1, B_2, C_1, C_2) were noted. The embryo development follows the Alysicarpus variation of the Onagrad type of Johansen (1950) or First Period, Series A. Megarchtype IV of Soueges and Crete (1952). In fact Sareen and Wadhwa (1981) could not go beyond Onagrad type. The structure of the testa agrees in general with the Papilionaceous type of Corner (1951). Thus it is confirmed that the observations of Sareen and Wadhwa (1981) on the embryology of *Alysicarpus vaginalis* are superficial and misleading.

Keywords: Embryology of angiosperms.

Introduction

The embryogeny of the Papilionaceae is full of interest. In this family, so well characterized by the structure of its flower and fruit, the degree of homogenity is apparently so great that the systematist hesitates in setting the limits of the various genera within the family. However, from the embryogenic point of view these genera can be as clearly distinguished as those of the Papaveraceae. The Papilionace has long been an object for embryological studies on account of considerable variation that exists in the mode of embryonal development so much so that even two different Megarchtypes may occur in the same species as is reported by Rau (1954) in *Desmodium laevigatum* (Hedysareae), Goursat (1969) in *Astragalus glycyphyllos* (Astragaleae) and *Baptisa australis* (Podalyrieae). However, Salgare (1973e, 74a, 97c) has observed three different Megarchtypes in *Phaseolus aconitifolius* (Phaseoleae), out of these three, the first two could be placed in Soueges and Crete (1952) embryogenic classification, but the third could not be placed in their system and seems to be a type by itself. In addition to the transverse division of the Oospore either vertical (Piperad type) or an obliquely transverse divisions were observed by Salgare (1975p) in *Sesbania aegyptiaca* (Galegeae). In *Alysicarpus vaginalis* (Hedysareae) six different Megarchtypes (A_1, A_2, B_1, B_2, C_1, C_2) were noted by Salgare (1986b). Two different Megarchtypes were noted by Salgare (unpublished) in *Phaseolus aureus* (Phaseoleae).

Results and Discussion

In *Alysicarpus vaginalis* the male archesporium is multi-cellular and hypodermal. The anther wall is four-layered and its development confirmed to the Dicotyledonous type. The tapetum is uni-nucleate, uni-seriate and secretory type. The endothecium forms fibrous thickenings at maturity. Simultaneous cytokinesis results in tetrahedral and isobilateral tetrads. Pollen grains are shed at bi-nucleate and bi-celled stage (1975d, 76d). Similar condition was also observed by Salgare in *Phaseolus aureus*

(1970, 73d, 75f, 86a), in *Phaseolus aconitifolius* (1974a, 75q, 76p, 97d), in *Dumasia villosa* (1975aa), *Cyamopsis psoralioides* (1975as), in *Sesbania aculeata* (1975ab, 76a, s), in *Sesbania aegyptiaca* (1976b, r). However, Sareen and Wadhwa (1981) reported uni-cellular archesporium, Monocotyledonous type of anther wall development and decussate microspore tetrads. They also failed to report three-nucleate pollen. Salgare (1975d, 76d) observed *in situ* germination of polloen in *Alysicarpus vaginalis*. It was the failure of Sareen and Wadhwa (1981) to report such an interesting observations. Salgare (1975d, 76d) stated that in *Alysicarpus vaginalis* the ovule is bitegmic, crassinucellate and campylotropous. The female archesporium which is uni- or bi-cellular is hypodermal in origin. A linear tetrad of megaspores is formed. The development of the megagametophyte is confirmed to the Polygonum type. Some abnormalities were observed during the development of the megagametophyte, indicating that the nuclear divisions in the megagametophyte are not always simultaneous resulting in three-, five- and six-nucleate megagametophytes. At one instance in the eight-nucleate an anomalous megagametophyte, polar nuclei were missing and two extra antipodals were found arranged in two series (3+2). Similar condition was also observed by Salgare in *Phaseolus aureus* (1970, 73d, 75f, ah, 76e, j, k, t, 78a, 80a, 86a, 97a, b, 2000), in *Phaseolus aconitifolius* (1974a, 75q, ac-ag, 76e-k, p, 77a, b, 78a, 80a, b, 81, 97a-c, 2000), in *Dumasia villosa* (1975v-aa, ai, aq, 76e-k, t, 77a, b, 78a, 80a, b, 97a, b, 2000), *Cyamopsis psoralioides* (1973b, 75a, m, n, ai, aq, as, 76e-h, j, k, q, t, 77a, 78a, 80a, b, 97a, b, 2000), in *Sesbania aculeata* (1973a, c, 75c, e, ab, aq, 76a, e-j, k, s, t, 77a, b, 78a, 80a, b, 97a, b, 2000), in *Sesbania aegyptiaca* (1974d, 75r-u, ai, aq, 76b, c, e-k, r, t, 77a, b, 78a, 80a, b, 97a, b, 2000) with a large number of very interesting anomalies. However, Sareen and Wadhwa (1981) were not aware of any type of anomaly in the megagametophyte.

In *Alysicarpus vaginalis* fertilization is porogamous. Salgare (1975b, d, o, aj, ar, 76n, 78b, 86b) stated that though double fertilization is a rule in *Alysicarpus vaginalis*, occasionally single fertilization that is syngamy occurred without triple fusion. Very often the zygotic nucleus was found to divide prior to the primary endosperm nucleus. The endosperm development follows the nuclear type. Sareen and Wadhwa (1981) were unaware of such anomalies. Salgare (1975b, d, ar, 76n, 86b) observed six different Megarchtypes (A_1, A_2, B_1, B_2, C_1, C_2) in *Alycarpus vaginalis*. The embryo development follows the Alysicarpus variation of the Onagrad type of Johansen (1950) or First Period, Series A, Megarchtype IV of Soueges and Crete (1952). In fact Sareen and Wadhwa (1981) could not go beyond Onagrad type. The structure of the testa agrees in general with the Papilionaceous type of Corner (1951). Thus it is confirmed that the observations of Sareen and Wadhwa (1981) on the embryology of *Alysicarpus vaginalis* are superficial and misleading.

Extensive work of Salgare (1973e, 74a, 75q, ac-ag, 76e-k, m, p, t, 77a, b, 78a, 80a, b, 81, 97a-d, 2000) on the embryology of *Phaseolus aconitifolius* proved that Bhasin (1971) and Deshpande and Bhasin (1974) were not aware of the fact that in addition to the uni-cellular male archesporium, bi-cellular archesporium and linear megaspore tetrad, in addition to T-shaped tetrads were also present. It was their failure to trace out the superimposed twin megagametophytes and superposed multiple megagametophytes and their further development into bisporic and trisporic development respectively. Bhasin (1971) and Deshpande and Bhasin (1974) also failed to trace out the endosperm haustorium and the development of the barrier tissue. In addition to the category A_2 and C_2 of Soueges and Crete (1952) Salgare (1973e, 97c) also recorded an additional tetrad of proembryoes which can fit in any of the categories of Soueges and Crete (1952) and forms the type by themself. This proves that the observations of Bhasin (1971) and Deshpande and Bhasin (1974) on the embryology of *Phaseolus aconitifolius* are superficial and misleading.

In reinvestigation of the embryology of *Cajanus cajan*, it was noted that there are three different types of megaspore tetrads *viz.* linear, T-shaped and a third one where the lower dyad member divides by a transverse wall, while meiosis II proceeds in the upper dyad member without cytokinesis. This third pattern cannot be accommodated under any of the Rembert's (1969) megaspore tetrad patterns and thus would form an independent patterns (Salgare, 1980c, 95). Roy (1933) had failed to report this type of tetrad in *Cajanus indicus* syn. of *Cajanus cajan*. As a result of the non-simultaneous nuclear divisions of the megagametophyte a three-nucleate megagametophyte was also noted in *Cajanus cajan* which was escaped from the observations of Roy (1933). Nine-nucleate an anomalous megagametophyte was reported by Roy (1933) in *Dolichos lablab*. However, it was the failure of Roy (1933) to decide the fate of an additional nucleus. It was Salgare (1975am, 80c) who proved that an additional nucleus in an anomalous megagametophyte of *Dolichos lablab* contributed to the formation of the secondary nucleus – resulting into triploid secondary nucleus in the Polygonum type of megagametophyte.

The ovule of *Phaseolus aureus* is bitegmic, crassinucellate and campylotropous. Though the outer integument is initiated later it grows faster and by itself alone forms the micropyle (Salgare, 1970, 73d, 75f, 76e, j, k, t, 78a, 80a, 86a, 97a, b, 2000; Salgare and Dnyansagar, 1971). However, George, George and Herr (1979) have stated that both the integuments are initiated simultaneously. This is an error due to their inability to get the earlier stages of integument development. The earliest stage which they have described (their Fig.14) is in fact, a more advanced stage and by no means the earliest. Hence a degree of confusion and misinterpretation has inadvertantly been produced. The inner integument consists of two layers throughout its development and the outer integument which is bi-layered in the beginning becomes thicker. In one case it was observed that both the integuments were of the two layers. Normally outer integument reaches at the top of the nucellus at the megaspore mother cell stage. But in some cases it has been observed that even at the dyad and tetrad stage both the integuments are creeping at the base of the nucellus. Such a variability in the nature and behaviour of the integuments in the same species of the Papilionaceae seems to be the first report. However, George, George and Herr (1979) were unaware of it. In addition to linear tetrads, T-shaped ones and an oblique T-shaped tetrad of megaspores were also noted by Salgare (1970, 73d, 75f, 76e, j, k, t, 78a, 80a, 86a, 97a, b, 2000) in *Phaseolus aureus*. George, George and Herr (1979) failed to note T-shaped and an oblique T-shaped tetrads which proved their superficial and misleading observations. Further they stated that the chalazal dyad cell divides unequally such that D (chalazal functional megaspore) is much larger than the a, b or c megaspore (their Figs. 19, 41, 42). Once again, from their Figures 19, 41, 42 it appears that they have mistaken a later stage for an earlier one, where the functional megaspore is considerably increased in size which accounts for their error of interpretation. So far there is no report of an unequal division of dyad amongst the Papilionaceae. Further an abnormal case was observed by Salgare (1970, 73d, 75f, 76e, t, 78a, 86a), where the megagametophyte was having an extra nucleus – 9-nucleate. George, George and Herr (1979) failed to take notice of such anomalies. With such a superficial observations they are comparing the development of ovule and megagametophyte in field-grown with the greenhouse-grown plants.

While monosporic development in megagametogenesis is the rule in Papilionaceae, bisporic development has occurred in *Lathyrus odoratus* (Jonsson, 1879-1880), in *Lupinus luteus* and *Lupinus polyphyllus* (Guignard, 1881), in *Laburnum anagyroides* (Rembert, 1966), in *Wisteria sinensis* (Rembert, 1967a) as well as in *Puereria lobata* (Rembert, 1969b), in *Canavalia ensiformis* (Salgare, 1975g-j, 76g, 77a, 80a, b, 97a, b, 2000), in *Canavalia gladiata* (Salgare, 1975l, 76g, 77a, 80a, 97a, b, 2000), in *Cyamopsis psoralioides* (Salgare, 1973b, 75m, as, 76g, q, t, 77a, 78a, 80a, b, 97a, b, 2000), in *Dumasia villosa* (Salgare, 1975z, aa, 76g, 77a, 80a, b, 97a, b, 2000), in *Phaseolus aconitifolius* (Salgare, 1973e, 74a, 75q, ad, 76i, m, p, t, 77a, b, 80a, b, 97a-c, 2000), in *Sesbania aculeata* (Salgare, 1975c, e, ab, 76a, i, 77b, 78a, 80a, b, 97a, b, 2000), in *Sesbania aegyptiaca* (Salgare, 1975r, s, 76b, c, g, h, i, r, 77b, 80a, b, 97a, b, 2000). It should be pointed out that all previous reports of bisporic development in Leguminosae have been challenged by Maheshwari (1955). Extensive work of Salgare make it very clear that the bisporic development in Leguminosae is a well established fact which invalid the challenge of Maheshwari (1955).

Salgare's (1970, 73a-e, 74a-d, 75a-as, 76a-t, 77a, b, 78a, b, 80a-c, 81, 86a, b, 95, 97a-d, 2000) outstanding contribution in the field of embryology of Papilionaceae is as: (1) Tendency towards bi-layered tapetum, (2) Bi-nucleate microspore mother cells, (3) Giant pollen grains, (4) Twin megagametophytes, (5) Juxtaposed twin megagametophytes, (6) Superposed twin megagametophytes, (7) Superimposed twin megagametophytes, (8) Superposed multiple megagametophytes, (9) Superposed multiple megagametophytes, (10) Superimposed superposed multiple megagametophytes, (11) Juxtaposed superposed multiple megagametophytes, (12) Non-simultaneous formation of antipodals and egg apparatus, (13) Occurrence of additional nuclei in antipodals, (14) Reduction in the number of antipodal cells, (15) Failure of the development of antipodals, (16) Megagametophytes with increased number of antipodals, (17) Failure of cell formation amongst antipodal nuclei, (18) Separation of antipodals from the main body of megagametophyte, (19) Megagametophytes with reversed polarity, (20) Egg with an additional nuclei, (21) Avortion of cell formation by egg nucleus, (22) Megagametophyte with suppression of egg, (23) Synergids with additional nuclei, (24) Occasional omission of synergid nuclei from cell formation, (25) Suppression of synergids, (26) Formation of secondary nucleus by more than two nuclei in Polygonum type of megagametophyte, (27) Megagametophyte without development of polar or secondary nucleus, (28) Eight new Megaspore Tetrad Patterns in Papilionaceae, (29) New Megagametophyte type – Trisporic Development, (30) Occasional occurrence of a single fertilization, (31) Prior division of zygotic nucleus instead of primary endosperm nucleus, (32) More than one type of embryo development in the same species, (33) Six different Megarchtypes in the same species and (34) New Megarchtypes. These are the first and only reports indicating that all the previous reports on the embryology of Papilionaceae are superficial and misleading.

References

Bhasin, R.K., 1971. Embryology of *Phaseolus aconitifolius*. In: *Proc. 58ᵗʰ Session Indian Sci. Congr.*, January 3–7, Bangalore Univ., Bangalore, Section, Abstract No. 156.

Corner, E.J.H., 1951. The leguminous seed. *Phytomorphology*, 1: 117–150.

Deshpande, P.K. and Bhasin, R.K. 1974. Embryological studies in *Phaseolus aconitifolius* Jacq. *Obs. Bot. Gaz.*, 135: 104–113.

George Glenda, P., George Ralph, A. and Herr, J.M. Jr., 1979. Comparative study of ovule and megagametophyte development in field-grown and greenhouse-grown plants of *Glycine max* and *Phaseolus aureus* (Papilionaceae). *Amer. J. Bot.*, 66: 1033–1043.

Goursat, Mazie-Jose, 1969. Researches sur L'embryogenic De Papilionaceae. *Ph.D. Thesis*, Faculty of Pharmacy. Series E. No. 190, Univ. De Paris.

Guignard, L., 1881. Recherches d'embryogenie vegetale Compree. 1. Legumineuses. *Ann. Sci. Nat. Bot.*, 12: 5–166.

Johansen, D.A., 1950. *Plant Embryology*. Waltham, Mass. U.S.A.

Jonsson, B., 1879-80. Om embryosackens utveckling hos Angiospermerna. Lunds Univ. Arsskr., 16: 1–86.

Maheshwari, S.C., 1955. The occurrence of bisporic embryo sacs in angiosperms: A critical review. *Phytomorphology*, 5: 67–99.

Rau, M.A., 1954. The development of embryo of *Cyamopsis*, *Desmodium* and *Lespedeza*, with a discussion on the position of the Papilionaceae in the system of embryogenic classification. *Phytomorphology*, 4: 418–430.

Rembert, D.H.Jr., 1966. Megasporogenesis in *Laburnum anagyroides* Medic. A case of bisporic development in Leguminosae. *Trans. Ky Acad. Sci.*, 27: 47–50.

Rembert, D.H.Jr., 1967a. Comparative megasporogenesis in Leguminosae – A Phylogenetic tool. *Ph.D. Diss.*, Univ. Kentucky, Lexington, U.S.A.

Rembert, D.H.Jr., 1967b. Development of the ovule and megagametophyte in *Wisteria sinensis*. *Bot. Gaz.*, 128: 223–229.

Rembert, D.H.Jr., 1969. Comparative megasporogenesis in Papilionaceae. *Amer. J. Bot.*, 56: 584–591.

Rembert, D.H.Jr., 1971. Phylogenetic significance of megaspore tetrad patterns in leguminosae. *Phytomorphology*, 21: 2–9.

Roy, B., 1933. Studies in the development of the female gametophyte in some leguminous crops in India. *Indian J. Agric. Sci.*, 3: 1098–1107.

Salgare, S.A., 1970. In: Embryological studies in *Phaseolus aureus* Roxb. *M.Sc. Thesis*, Univ. Bombay.

Salgare, S.A., 1973a. On the megagametophyte of *Sesbania aculeata*. In: *Proc. 60ᵗʰ Session Indian Sci. Congr.*, Botany Section 3: 332.

Salgare, S.A., 1973b. On the megagametophyte of *Cyamopsis psoralioides*. In: *Proc. 60ᵗʰ Session Indian Sci. Congr.*, Botany Section 3: 332–333.

Salgare, S.A., 1973c. II. On the megagametophyte of *Sesbania aculeata* Poir. *Sci. and Cult.*, 39: 309–311.

Salgare, S.A., 1973d. A note on the embryology of *Phaseolus aureus* Roxb. *Curr. Sci.*, 42: 869–871.

Salgare, S.A., 1973e. On the early embryogeny of *Phaseolus aconitifolius*. *Sci. and Cult.*, 39: 315–316.

Salgare, S.A., 1974a. Embryology of *Phaseolus aconitifolius*, with a discussion on the position of the Phaseoleae in the system of embryogenic classification. In: *Proc. 61ˢᵗ Session Indian Sci. Congr.*, Botany Section, Abstract No. 2.

Salgare, S.A., 1974b. Development of the seed of *Phaseolus aureus* Roxb. *J. Univ. Bom.*, 43: 87–98.

Salgare, S.A., 1974c. Status of the Papilionaceae. *J. Biol. Sci.*, 17: 82–85.

Salgare, S.A., 1974d. The megagametophyte of *Sesbania aegyptiaca* Poir. I. *J. Biol. Sci.*, 17: 108–110.

Salgare, S.A., 1975a. Trisporic development in *Cyamopsis psoralioides* DC.In: *Proc. 62ⁿᵈ Session Indian Sci. Congr.*, January 3–7, Botany Section, Abstract No. 116.

Salgare, S.A., 1975b. On the early embryogeny of *Alysicarpus vaginalis* DC. In: *Proc. 62ⁿᵈ Session Indian Sci. Congr.*, January 3–7, Botany Section, Abstract No. 117.

Salgare, S.A., 1975c. III. On the megagametophyte of *Sesbania aculeata* Poir. *Sci. and Cult.*, 41: 166–167.

Salgare, S.A., 1975d. The embryology of *Alysicarpus vaginalis* DC. *Biovigyanam*, 1: 73–74.

Salgare, S.A., 1975e. IV. On the megagametophyte of *Sesbania aculeata* Poir. *Sci. and Cult.*, 41: 172.

Salgare, S.A., 1975f. Gametophytes of *Phaseolus aureus* Roxb. *J. Indian Bios. Asso.*, 1: 1–6.

Salgare, S.A., 1975g. I. On the megagametophyte of *Canavalia ensiformis* DC. *J. Indian Bios. Asso.*, 1: 7–9.

Salgare, S.A., 1975h. II. On the megagametophyte of *Canavalia ensiformis* DC. *J. Indian Bios. Asso.*, 1: 9–11.

Salgare, S.A., 1975i. III. On the megagametophyte of *Canavalia ensiformis* DC. *J. Indian Bios. Asso.*, 1: 12–13.

Salgare, S.A., 1975j. IV. On the megagametophyte of *Canavalia ensiformis* DC. *J. Indian Bios. Asso.*, 1: 14–15.

Salgare, S.A., 1975k. V. On the megagametophyte of *Canavalia ensiformis* DC. *J. Indian Bios. Asso.*, 1: 15–16.

Salgare, S.A., 1975l. I. On the megagametophyte of *Canavalia gladiata* DC. *J. Indian Bios. Asso.*, 1: 16–18.

Salgare, S.A., 1975m. Bisporic development in *Cyamopsis psoralioides* DC. *J. Indian Bios. Asso.*, 1: 18–22.

Salgare, S.A., 1975n.Trisporic development in *Cyamopsis psoralioides* DC. *J. Indian Bios. Asso.*, 1: 22–25.

Salgare, S.A. 1975o. Syngamy with occasional single fertilization in *Alysicarpus vaginalis* DC. *J. Indian Bios. Asso.*, 1: 25–27.

Salgare, S.A., 1975p. On the early embryogeny of *Sesbania aegyptiaca* Poir. *J. Indian Bios. Asso.*, 1: 27–30.

Salgare, S.A., 1975q. Gametophytes of *Phaseolus aconitifolius* Jacquin. Obs. *J. Indian Bios. Asso.*, 1: 35–50.

Salgare, S.A., 1975r. III. On the megagametophyte of *Sesbania aegyptiaca* Poir. *J. Indian Bios. Ass.*, 1: 66–67.

Salgare, S.A., 1975s. IV. On the megagametophyte of *Sesbania aegyptiaca* Poir. *J. Indian Bios. Ass.*, 1: 67–68.

Salgare, S.A., 1975t. V. On the megagametophyte of *Sesbania aegyptiaca* Poir. *J. Indian Bios. Ass.*, 1: 68–70.

Salgare, S.A., 1975u. VI. On the megagametophyte of *Sesbania aegyptiaca* Poir. *J. Indian Bios. Ass.*, 1: 70–75.

Salgare, S.A., 1975v. On the megasporogenesis of *Dumasia villosa* DC. *J. Indian Bios. Ass.*, 1: 76–80.

Salgare, S.A., 1975w. I. On the megagametophyte of *Dumasia villosa* DC. *J. Indian Bios. Ass.*, 1: 80–81.

Salgare, S.A., 1975x. II. On the megagametophyte of *Dumasia villosa* DC. *J. Indian Bios. Ass.*, 1: 81–85.

Salgare, S.A., 1975y. III. On the megagametophyte of *Dumasia villosa* DC. *J. Indian Bios. Ass.*, 1: 85–89.

Salgare, S.A., 1975z. IV. On the megagametophyte of *Dumasia villosa* DC. *J. Indian Bios. Ass.*, 1: 89–93.

Salgare, S.A., 1975aa. Gametophytes of *Dumasia villosa* DC. *J. Indian Bios. Ass.*, 1: 97–116.

Salgare, S.A., 1975ab. Gametophytes of *Sesbania aculeata* Poir. *J. Indian Bios. Ass.*, 1: 143–163.

Salgare, S.A., 1975ac. I. On the megagametophyte of *Phaseolus aconitifolius* Jacquin, Obs. *J. Indian Bios. Ass.*, 1: 164–165.

Salgare, S.A., 1975ad. II. On the megagametophyte of *Phaseolus aconitifolius* Jacquin, Obs. *J. Indian Bios. Ass.*, 1: 165–167.

Salgare, S.A., 1975ae. III. On the megagametophyte of *Phaseolus aconitifolius* Jacquin, Obs. *J. Indian Bios. Ass.*, 1: 167–169.

Salgare, S.A., 1975af. IV. On the megagametophyte of *Phaseolus aconitifolius* Jacquin, Obs. *J. Indian Bios. Ass.*, 1: 169–172

Salgare, S.A., 1975ag. V. On the megagametophyte of *Phaseolus aconitifolius* Jacquin, Obs. *J. Indian Bios. Ass.*, 1: 172–176.

Salgare, S.A., 1975ah. Giant pollen grains of *Phaseolus aureus* Roxb. *J. Indian Bios. Ass.*, 1: 176–177.

Salgare, S.A., 1975ai. Twin megaspore tetrads in the Papilionaceae. *J. Indian Bios. Asso.*, 1: 177–180.

Salgare, S.A., 1975aj. I. Occasional occurrence of the prior division of the zygotic nucleus in the Papilionaceae. *J. Inidan Bios. Asso.*, 1: 180–181.

Salgare, S.A., 1975ak. Endosperm haustoria and formation of the barrier tissue in *Phaseolus aureus* Roxb. *J. Indian Bios. Ass.*, 1: 181–182.

Salgare, S.A., 1975al. On the megasporogenesis of *Dolichos lablab* Linn. *J. Indian Bios. Ass.*, 1: 182–185.

Salgare, S.A., 1975am. I. On the megagametophyte of *Dolichos lablab* Linn. *J. Indian Bios. Ass.*, 1: 185–187.

Salgare, S.A., 1975an. II. On the megagametophyte of *Dolichos lablab* Linn. *J. Indian Bios. Ass.*, 1: 187–189.

Salgare, S.A., 1975ao. I. On the megagametophyte of *Alysicarpus vaginalis* DC. *J. Indian Bios. Ass.*, 1: 189–192.

Salgare, S.A., 1975ap. II. On the megagametophyte of *Alysicarpus vaginalis* DC. *J. Indian Bios. Ass.*, 1: 192–194.

Salgare, S.A., 1975aq. I. Occasional omission of the second meiotic division in one of the dyad members of the Papilionaceae. *J. Indian Bios. Ass.*, 1: 194–198.

Salgare, S.A., 1975ar. Fertilization and early embryogeny of *Alysicarpus vaginalis* Dc. *J. Indian Bios. Ass.*, 1: 198–205.

Salgare, S.A., 1975as. Male and female gametophytes of *Cyamopsis psoralioides* DC. *Biovigyanam*, 1: 173–181.

Salgare, S.A., 1976a. Embryology of *Sesbania aculeata* Poir. In: *Proc. 63ʳᵈ Session Indian Sci. Congr*, Botany Section, Abstract No. 102.

Salgare, S.A., 1976b. Embryology of *Sesbania aegyptiaca* Poir. In: *Proc. 63ʳᵈ Session Indian Sci. Congr*, Botany Section, Abstract No. 103.

Salgare, S.A., 1976c. Gametophytes of *Sesbania aegyptiaca* Poir. *J. Indian Bios. Ass.*, 2: 1–14.

Salgare, S.A., 1976d. Gametophytes of *Alysicarpus vaginalis* DC. *J. Indian Bios. Ass.*, 2: 15–33.

Salgare, S.A., 1976e. I. On the megagametophyte of Papilionaceae. *J. Indian Bios. Ass.*, 2: 43–48.

Salgare, S.A., 1976f. II. On the megagametophyte of Papilionaceae. *J. Indian Bios. Ass.*, 2: 53–56.

Salgare, S.A.,1976g. I. Superposed twin megagametophytes in Papilionaceae. *J. Indian Bios. Ass.*, 2: 56–63.

Salgare, S.A., 1976h. On the megagametophyte of Papilionaaceae – I. The synergids. *J. Indian Bios. Ass.*, 2: 63–70.

Salgare, S.A., 1976i. I. Superimposed twin megagametophytes in Papilionaceae. *J. Indian Bios. Ass.*, 2: 70–78.

Salgare, S.A. 1976j. On the megagametophyte of Papilionaceae –I. The antipodals. *J. Indian Bios. Ass.*, 2: 107–121.

Salgare, S.A., 1976k. On the megagametophyte of Papilionaceae – Non– simultaneous formation of the antipodals and egg apparatus. *J. Indian Bios. Ass.*, 2: 125–137.

Salgare, S.A., 1976l. I. Occasional occurrence of a single fertilization in Papilionaceae. *J. Indian Bios. Ass.*, 2: 142–152.

Salgare, S.A., 1976m. Development of the seed of *Phaseolus aconitifolius* Jacquin, Obs. *J. Indian Bios. Ass.*, 2: 170–178.

Salgare, S.A., 1976n. Fertilization and development of the seed of *Alysicarpus vaginalis* DC. *J. Indian Bios. Ass.*, 2: 179–185.

Salgare, S.A., 1976o. The development of endosperm in *Phaseolus aureus* Roxb. *J. Indian Bios. Ass.*, 2: 232–234.

Salgare, S.A., 1976p. Embryology of *Phaseolus aconitifolius* Jacquin, Obs. *J. Indian Bios. Ass.*, 2: 234–239.

Salgare, S.A., 1976q. On the megasporogenesis of *Cyamopsis psoralioides* DC. *J. Indian Bios. Ass.*, 2: 239–243.

Salgare, S.A., 1976r. Embryology of *Sesbania aegyptiaca* Poir. *J. Indian Bios. Ass.*, 2: 243–246.

Salgare, S.A., 1976s. Embryology of *Sesbania aculeata* Poir. *J. Indian Bios. Ass.*, 2: 263–266.

Salgare, S.A., 1976t. On the megagametophyte of Papilionaceae. Proc. All India UGC sponsored Seminar – Recent trends and contact between Cytogenetics, Embryology and Morphology, held on October 15–17, 1976 at Nagpur Univ., Nagpur. Abstract No. 46.

Salgare, S.A., 1977a. Superposed twin megagametophytes in Papilionaceae. In: *Proc. 64th Session Indian Sci. Congr*, Botany Section, Abstract No. 128.

Salgare, S.A., 1977b. Superimposed twin megagametophytes in Papilionaceae. In: *Proc. 64th Session Indian Sci. Congr*, Botany Section, Abstract No. 129.

Salgare, S.A., 1978a. On the megagametophyte of Papilionaceae–I, Non-simultaneous formation of the antipodals and egg apparatus. In: *Proc. 65th Session Indian Sci. Congr.*, held on January 3–7, Botany Section, Abstract No.196.

Salgare, S.A., 1978b. Occasional occurrence of a single fertilization in Papilionaceae. In: *Proc. 65th Session Indian Sci. Congr.*, on January 3–7, Botany Section, Abstract No. 197.

Salgare, S.A., 1980a. I. A Criticism of Rembert's papers entitled, 'Comparative Megasporogenesis in Papilionaceae and Phylogenetic Significance of Megaspore Tetrad Patterns in Leguminales.' In: *Proc. 2nd All India Symp. Life Sci. Proceeded with Ann. Conf. Indian Soc. Life Sci.*, held on March 9–11 Govt. Inst. of Sci., Nagpur. Abstract No. 43.

Salgare, S.A., 1980b. Bisporic development in Papilionaceae–Challenge to the Hypothesis of Maheshwari (1955). In: *Proc. 2nd All India Symp. Life Sci. Proceeded with Ann. Conf. Indian Soc. Life Sci.*, held on March 9–11, Govt. Inst. of Sci., Nagpur. Abstract No. 44.

Salgare, S.A., 1980c. Challenge to the findings of roy (1933). In: *Proc. 2nd All India Symp. Life Sci. Proceeded with Ann. Conf. Indian Soc. Life Sci.*, held on March 9–11, Govt. Inst. of Sci., Nagpur, Abstract No. 45.

Salgare, S.A., 1981. Status of the papilionaceae with special reference to Phaseoleae. In: *Proc. 68th Session Indian Sci. Congr.*, January 3–7, Botany Section, Abstract No.157.

Salgare, S.A., 1986a. Embryological studies in *Phaseolus aureus*. In: *Proc. Special Indian Geophytological Conf.*, held on November 27–29, at Univ. of Poona. Abstract No. 43.

Salgare, S.A., 1986b. Embryological studies in Papilionaceae: The genus *Alysicarpus* Neck Ex Desv. In: *Proc. Special Indian Geophytological Conf.*, November 27–29, at Univ. of Poona, Abstract No. 44.

Salgare, S.A., 1995. Reinvestigation of the embryology of *Cajanus cajan* DC. In: *Proc. Nat. Symp. on Recent Advances in Biosciences*, November 3–5 at Maharshi Dayanand Univ., Rohtak. Abstract No. 238.

Salgare, S.A. 1997a. Megaspore tetrad patterns in Papilionaceae: A critical review. *Flora and Fauna*, 3: 66–70.

Salgare, S.A., 1997b. Present status of megaspore tetrad pattern of Papilionaceae: Recent trends in life sci. In: *Proceeded with 14th Nat. Symp. on Indian Soc. of Life Sci.*, October 23–25, Pubjabi Univ., Patiala. Abstract No. 91.

Salgare, S.A., 1997c. A New Megarchtype. In: *Recent Trends in Life Sci. Proceeded with 14th Nat. Symp. on Indian Soc. of Life Sci.*, October 23–25, Pubjabi Univ., Patiala. Abstract No. 92.

Salgare, S.A., 1997d. Present status of the embryology of *Phaseolus aconitifolius* Jacquin, Obs. In: *Recent Trends in Life Sci. Proceeded with 14th Nat. Symp. on Indian Soc. of Life Sci.,* October 23–25, Pubjabi Univ., Patiala. Abstract No. 94.

Salgare, S.A. 2000. Megaspore tetrad patterns in Papilionaceae: A critical review. In: *Proc. 87th Session Indian Sci. Congr.,* January 3–7, Univ. Pune, Pune, Botany Section, Abstract No.73.

Salgare, S.A., and V.R. Dnyansagar, 1971. Embryology of *Phaseolus aureus.* In: *Proc. 58th Session Indian Sci. Congr.,* January 3–7, Bangalore Univ., Bangalore. Botany Section, Abstract No. 155.

Sareen, T.S. and K. Wadhwa, 1981. Embryological studies in Papilionaceae: The genus *Alysicarpus* neck. In: *Proc. 68th Session Indian Sci. Congr.,* January 3–7, Botany Section, Abstract No. 153.

Soueges, R. and Crete, P., 1952. Les acquisitions les plus recentes de l'embryogenie des Angiospermes (1947–1951) *Ann. Biol.,* 28: 9–45.

Chapter 16

Ayurvedic Pharmacopoeial Plant Resources of Capparidaceae

Anand Prakash, Md. Nizamuddin Ansari,
Chandan Kr. Singh and Md. Naseem

Tissue Culture Laboratory, B.R.A. Bihar University,
Muzaffarpur, Bihar

ABSTRACT

The prevalent practice of herbal remedies have descended down from generation to generation and includes the cure from simple ailments to the most complicated ones like snake-bite, stomach disorders, dog-bite, bone fracture, diabetes and even birth control. India has old tradition of herbal medicines. Most of the drugs are obtained from plants collected in nature. It was found that the person practising this art does not reveal his knowledge to others. There are instances when the valuable information has been buried with the dying persons.

No concerted efforts have so far been made regarding these drug plants for their proper identification, cultivation, chemical constituents and their uses against various diseases.

In this background, the present study on these plants *viz. Cleome gynandra, Cleome viscosa, Capparis spinosa* and *Crataeva religiosa* of family Capparidaceae is highly desirable and was made at preliminary level. It is desirable that steps to be undertaken for their conservation, protection, propagation and systematic and scientific exploitation for meeting out the future need.

Keywords: Pharmacopoeial, Efficacy, Gynandrophore, Wasteland, Medicinal plant.

Introduction

Muzaffarpur district is situated in north Bihar and lies 26.8° north latitude and 22.5° east latitude covering an area of 3172 square kilometer with an elevation of 84m, mean sea level. This district is situated in the south of tarai belt of Nepal. Diverse group of medicinal plants are found to grow in this area. Medicinal plants are in demand since the beginning of human civilization (Chopra *et al.*, 1956; Kaushik, 1988) and various plant products features prominently in traditional therapeutics. Over 25 per cent of all drugs dispensed world over include plant constituents (Mehta, 1989; Ahuja, 1994; Chatterjee and Prakashi, 1997). The information on the use of medicinal plants is scattered and most of it is found in books and periodicals, many of which are out of print and are not available even in large libraries. One of the greatest difficulties confronting the research workers is the paucity of authentic information on the identity, habitat and use of medicinal plants (Chopra *et al.*, 1956; Agarwal, 1986).

In the present investigation, preliminary studies were made on some medicinal plants of family Capparidaceae. There is no systematic cultivation and proper studies on these plants with respect to their biochemical constituents, medicinal uses and mass propagation. Due to ever increase in demand for food, fodder, medicines, population explosion and ruthless exploitation of plants for commercial purposes, there is rapid depletion of natural flora including medicinal plants (Ahuja, 1994; Tandon, 1994). No concerted efforts have so far been made regarding these drug plants for their proper cultivation, chemical constituents and their specific uses against diseases. Keeping in view the above facts, the present studies on some plants of *Capparidaceae viz. Cleome gynandra, Cleome viscosa, Capparis spinosa* and *Crataeva religiosa* were made. Since the members of *Capparidaceae* are well known for their officinal use, this study would be highly beneficial for general mass particularly rural and tribal people and the commercial organizations which deal in herbal drugs.

Materials and Methods

During the course of present studies, surveys were conducted twice in a year (February–March and September–October) in the urban and rural areas of Muzaffarpur to know the distribution pattern of plants under investigation in natural condition. These plants were identified by the help of local people and skilled persons dealing with herbs, further the taxonomic identification of plants under investigation was confirmed with the help of monograph entitled *"The Botany of Bihar and Orissa"* by H.H. Hains (Vol. 1, 1961). Experimental plants were collected, identified and transferred into herbarium sheets.

Medicinal uses and active constituents of these plants were collected from literature and local people practicing in herbal system of medicines. Regarding the use of experimental plants for the treatment of various ailments, data and protocols were collected from experts of herbal plants and local people. Lack of written information and reluctance of the practitioners to divulge their practice have unfortunately deprived the present generation of the very important and useful experiences and knowledge of our ancients in the field of medico-botany. Hence, the need of present study on proper identification and uses of medicinal plants is highly desirable.

Authors have tried and tested the effect of root decoction of *Cleome gynandra* on kids against cold – prone fever and broncho-pneumonia. Out of various members of *Cappridaceae*, studies only on four plants *viz. C. gynandra, C. viscosa. C. spinosa* and *C. religiosa* have been made in the present investigation.

Results and Discussion

In India, 7 genera of family *Capparidaceae* have been found in western and southern India and a few species grow in tropical Himalayas. These are: *Cleome, Capparis, Crataeva, Cadaba, Maerua, Roydsia* and *Gynandropsis*. Out of 7 known genera in India, only four species *viz. Cleome gynandra, Cleome viscosa, Capparis spinosa* and *Crataeva religiosa* are reported in Muzaffarpur with restricted distribution.

Plants are described giving local and vernacular names, brief description, distribution in Muzaffarpur, parts used and their medicinal uses.

Cleome gynandra L. (Fam: Capparidaceae, now in Cleomaceae)

Syn. Gynandropsis gynandra (Linn.)Briq.

Local Names

Hul-hul, Gandhuli, Sada hurhuria, Marang charmani.

A common annual herb measuring about 1-4 feet in height with pungent odour, stem green, cylindrical, branched and sticky with glandular hairs. Leaves pentafoliate and leaflets radiate from the tip of the stalk, inflorescence corymbose terminal raceme with many flowers and elongated fruits. Flowers are bisexual, hypogynous and white with distinct androphore and gynophore collectively known as gynandrophore – a unique floral character, petals are four, free with distinct claw and limb. Fruit is silique like linear capsule with numerous reniform, brown or black seeds attached parietally.

Distribution

Common in rainy season along road sides, garbage deposits, railway track and in barren lands, widely distributed in tropical parts in India.

Parts Used

Roots, leaves and seeds.

Local Uses

The root decoction is much effective against cold-prone fever and broncho-pneumonia. It was tested on 50 kids and was found that gently warm root decoction with 1 to 2 black pepper was highly effective against cold-prone fever and broncho-pneumonia (Naseem, 1990). This treatment is practiced in the country side by the poor and middle class people. Leaves are used to prevent the pus and boils and juice of the leaves mixed with oils is applied by local people for the relief of otalgia and catarrhal. Leaves also had anti-inflammatory properties. This plant also has insecticidal and repellent characteristics.

Seeds of this plants are anthelmintic and rubefacient and given internally for the expulsion of roundworms. Seeds mixed with oils are used as vermicide and in dressing the hair killing lice. Tribals even use this plant in snake-bite and scorpion-sting.

Cleome viscosa L. (Fam: Capparidaceae, now placed in Cleomaceae)

Local Names
Arkakanta, Hul-hul, Hurhur, Hur-huria

An erect, hairy annual herb with yellow flowers disposed on a long raceme and hermaphrodite, sticky and glandular herb with a strong penetrating odour, superior ovary with parietal placentation, gynophore absent. Slightly redish reniform seeds with pungent odour found in linear capsule fruits.

Distribution
Commonly found in rainy season along the roadside, garbage deposits and in waste land. It is abundant in cultivated fields especially in chaur (Naseem, 1990; Sharma, 2002).

Parts Used
Roots, leaves and seeds.

Local Uses
Leaves are applied externally for the cure of wounds and ulcers. The juice of leaves mixed with ghee is effective against inflammation in the middle ear. Seeds are used as carminative and anthelmintic.

Capparis spinosa L. (Fam: Capparidaceae)

Local Names
Kabra, kabar and capper plant.

A diffused prostate polymorphic shrub with variable leaves, hooked stipules, pedicillate and axillary flowers with white petals and purple filaments, long gynophore with ribbed fruits, globose and brown seeds.

Distribution
Commonly found on rocky places and road and railway sides in Muzaffarpur locality. Distribution of this plant is not frequent.

Parts Used
Bark of stem and root, leaves, flower buds, fruits.

Local Uses
This plant is a medicinally important taxon of family Capparidaceae and is credited with anti-tubercular property.

Crataeva religiosa Forst. F. var. *nurvala* (Fam: Capparidaceae)

Local Names
Varuna, varun, kumarak.

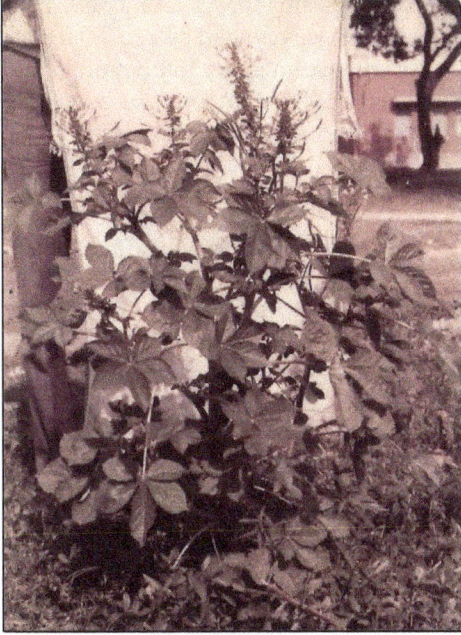

Figure 16.1: *Cleome gynandra* L.

Figure 16.2: *Cleome viscosa* L.

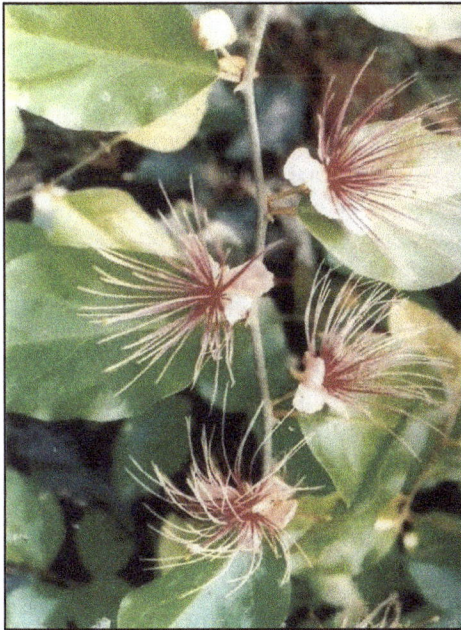

Figure 16.3: *Capparis spinosa* L.

Figure 16.4: *Crataeva religiosa* L.

A profusely branched tree upto 25-30 ft in height, stem is green, branched, hairy and solid. A crown like look of tree with simple leaves, leaves shed before flowering. Flowers are large, bisexual, actinomorphic corymbose-raceme, hypogynous and complete, flowers bloom in the evening, gynophore present, ovary is bicarpellary, syncarpous and elevated on a long stalk called gynophore, flower is highly attractive and beautiful.

Distribution
Crataeva is a tropical tree and is common throughout India. In Muzaffarpur, this plant is not reported to be found. A few plants are found growing in campus of RAU, Pusa (Samastipur) and Sanskrit Mahavidyalaya, Muzaffarpur.

Parts Used
Bark of root and stem.

Local Uses
Plants are used as herbal drugs for the cure of kidney and bladder problems. It is known as drug of chronic for treating urinary disorders. Stem and root barks are used against chronic mucous. Its key uses are in lithotropic and urinary calculi. As this plant contains triterpens alkaloid, it has established anti-inflammatory properties, potentially helpful for arthritis and hepatitis.

In Sanskrit, the medicinal uses of varuna have been described as follows:

ßo: .kks oj .k% lsrqfrdr'kkdP%dqekjdPA

o: .k% fir~ryks Hksrh'ys ed`Bk'eek:rku~ AA 65 AA

fu_fÍ re/kqjfÍrdP%d`Pksks:Kdks y?kdP%AÞ 66

In this back ground, indigenous drugs may be expected to be of value. It is desirable that efficacy of the herbal products may be enhanced by use of proper biotechnological technique (Ahuja, 1994; Naseem and Jha, 1994, 1997; Tandon, 1994; Singh, 2007; Najafi *et al.*, 2008; Thind *et al.*, 2008). Data on medicinal uses of these plants were gathered from herbal experts and local and tribal people. It is highly desirable that thorough studies be made on these medicinal plants with respect to their active constituents, systematic cultivation and conservation. To achieve this goal, large pharmaceutical houses like Dabur, Hamdard, Himalaya and other drug house should come forward to take this challenge and efforts should be made for proper formulation of drugs from these plant source which are wonderful gift of the nature (Ansari, 2007; Prakash, 2007). These plant products will solve the problem of millions of people especially kids suffering from diseases like broncho-pneumonia, liver-disorder, cough and fever (Naseem, 1990; Sharma, 2002).

Acknowledgement
The authors are grateful to Dr. S. Kumar, Professor and Head, University Department of Botany, B. R. A. Bihar University, Muzaffarpur for encouragement.

References

Ahuja, P.S., 1994. Role of plant tissue culture in the improvement of medicinal and aromatic plants. In: *Proceedings of XVII Plant Tissue Culture Conference*, BHU, Varanasi, p. 1.

Ansari, M.N., 2007. Studies on some antidiabetic drug plants. *M. Phil Dissertation*, Alagappa University, Karaikudi, India.

Chatterjee, A. and Prakashi, S.C., 1997. *The Treatise on Indian Medicinal Plants,* Vol. 5. National Institute of Science Communication, New Delhi.

Chopra, R.N., Nayar, S.L. and Chopra, I.C., 1956. *Glossary of Indian Medicinal Plants.* CSIR Publication, New Delhi.

Haines, H.H., 1961. *The Botany of Bihar and Orissa*. Botanical Survey of India. Sri Gouranga Press Pvt. Ltd., Calcutta, 1: 30–32.

Kaushik, P., 1988. *Indigenous Medicinal Plants*. Today and Tomorrow's Printers and Publishers , New Delhi.

Mehta, A.R., 1989. Some recent developments in *in vitro* research of plant products. In: *Proceedings of XIII Plant Tissue Culture Conference;* NEHU, Shillong, p. 3.

Najafi, S.H., Shitole, M.G. and Deokule, S.S., 2008. Numerical analysis based on morphological and histological characters of some medicinally important plants of the tribe Marsdenieae (*Asclepiadaceae*). *Phytomorphology*, p. 581.

Naseem, M., 1990. Studies on differentiation and organogenesis in tissue culture of some plants (*Gynadropsis pentaphylla* and *Cleome viscosa* L.) of medicinal importance. *Ph.D. Thesis*, BRA Bihar University, Muzaffarpur, India.

Naseem, M. and Jha, K.K., 1994a. Differentiation and regeneration in *Cleome* leaves cultured *in vitro. Egypt J. Bot.,* 34: 37–49.

Naseem, M. and Jha, K.K., 1997. Rapid clonal multiplication of *Cleome gynandra* DC.

Prakash, A., 2007. Studies on some medicinal plants of *Capparidaceae*. M. Phil tissue culture. *Phytomorphology,* 47: 405–411.

Sharma, S.N., 2003. Biochemical studies on tissue cultures of *Gyanandropsis gynandra* L. Briq. and *Cleome viscosa* L. *Ph.D Thesis*, B.R.A.Bihar University, Muzaffarpur, India.

Singh, J.P., 2007. *In vitro* propagation of a medicinal plant of solanaceae. *M. Phil Dissertation*, Alagappa University, Karaikudi, India.

Tandon, P., 1994. Role of tissue culture in plant conservation. In: *Proceedings of XVII Annual Plant Tissue Culture Conference*, BHU, Varanasi, India, pp. 41.

Thind, S.K., Jain, N. and Gosal, S.S., 2008. Micropropagation of *Aloe vera* L. and estimation of potentially active secondary constituents. *Phytomorphology*, 58: 65– 72.

Index